The Human Equation

Cosmogenesis + Biogenesis + Anthropogenesis
= Christogenesis

Roger Skrenes

En Route Books and Media, LLC
Saint Louis, MO

⊕ENROUTE
Make the time

En Route Books and Media, LLC

5705 Rhodes Avenue

St. Louis, MO 63109

Contact us at

contactus@enroutebooksandmedia.com

Cover credit: Sebastian Mahfood

Copyright © 2022 Roger Skrenes

ISBN-13: 978-1-956715-22-4

Library of Congress Control Number available at

https://catalog.loc.gov/vwebv/ui/en_US/

htdocs/help/numbers.html

TABLE OF CONTENTS

INTRODUCTION

"The Human Equation"

This book is an attempt to trace the footprints of the Creator since the event of the Big Bang. It is a snapshot in one volume of the movement and work of God in the world of our experience.

"So shall My Word be, that goes out of My mouth; it shall not return to Me void, but it shall do that which I please; and it shall prosper in what I sent it [to do]."

Isaiah 55:11

The first section of this book ("Cosmogenesis") traces the origin of matter, of stars, and of our own solar system. The second section ("Biogenesis") traces the origins of living things on this planet, plants and of animals, the sources of our food and of our bodies. The third section ("Anthropogenesis") explores the arrival of humans upon the earth.

The fourth and final section of this book is entitled "Christogenesis." It is the heart of the book. It is an attempt to document the work of humans on this planet and to show how this work points to our certain home in Heaven, also known as the Kingdom of God.

"The ages were put in order by the Word of God."

Hebrews 11:3

The first part of this fourth section will examine the history of medicine, whereby humans are raised from the dead – a sign of Christ's resurrection and of our own.

The second part of this fourth section will trace the history of transportation that is making mankind less bound in space. Such freedom of movement is one of the characteristics of life in Heaven. This freedom also mirrors one of the divine attributes of God, namely his "omnipresence" within the universe.

"Where shall I flee from your presence?
If I ascend up into heaven, you are there.
If I make my bed in Sheol [underground],
you are [also] there!"

Psalm 139:7

The third part of this fourth section contains an outline history of communications. In the area of communications, humans are becoming less bound both in space and in time. Time contraction is a reflection of the God who has no time, or is not in time. God is "eternal." Such a form of everlasting life has also been promised to his true followers.

"My sheep hear my voice. I know them, and they follow me. I give them eternal life and they will never perish."

John 10: 27-28

The last part of the fourth section examines the role of the factory in modern life. It will focus on various kinds of factory production as a means of creating a "New World" – which is a sign of our final home in Heaven.

"What no eye has seen, nor ear heard, nor the human heart conceived, what God has prepared for those who love him."

1 Corinthians 2:9

The four major sections of this book detail the various areas of God's presence and of his work in our world. They especially illustrate the divine face of evolution since the beginnings of Creation. This can be expressed by using a memory device or equation: C+B+A = C.

Cosmogenesis + Biogenesis + Anthropogenesis
= Christogenesis

COSMOGENESIS

"Cosmogenesis," as a word, involves at least one important distinction. That distinction is that "creation" and "evolution" are not identical realities. "Creation" refers to the fact that God has brought this world into existence from nothing. Evolution is what happens to matter <u>after it is created.</u> To evolve, the world first has to exist. From nothing comes nothing. Evolution requires that the Creator brings the material world into existence. In other words, evolution requires God.

"The heavens and the earth … God did not make them out of existing things …"

2 Maccabees 7:28

"The ages were put in order by the Word of God, so that what is visible [in this world] has not come from what is seen." [literal English from the NT Greek]

Hebrews 11:3

"For the invisible things of him [God] are clearly seen from the creation of the world, being understood by the things that are made …"

Romans 1:20

"Is anything too hard for Yahweh [God the Father]?"

Genesis 18:14

"Nothing will be impossible with God."

Luke 1:37

It was God, therefore, the divine architect, who brought the plan for our world into existence. Then, he directed it, in an unseen way, through secondary causes. Evolution is an expression of these unseen directives or causes of God over time.

GOD

1. God the "First Cause"

There are many causes of things in the world. Nothing in the world, however, is the cause of its own existence. If that were so, the thing (or person) would have had to exist before it came into existence – which is impossible. If the first thing in a long series of causes does not exist, nothing in that series could exist. Furthermore, if a series extended infinitely back into the past without a First Cause, there would be nothing existing today. From nothing, nothing can come into existence. However, there are many things in existence today! Therefore, there was certainly a First Cause to begin our world. This First Cause is God, who himself is uncaused.

2. God the "Unmoved Mover"

There are many kinds of motion. A growing plant is in motion. A college student studying for an exam is in motion. An older person advancing in age is in motion. In fact, everything in the universe is involved in some kind of motion. Furthermore, nothing in this world moves solely of and by itself. Everything is helped along in some way or ways by other persons or other outside forces. Things move when a potential (or possible) motion becomes an actual (or happening) motion. Everything in motion must *initially be* moved by something or someone other than itself. Furthermore, the causes of such motion cannot go on as a long series forever. At the beginning, there must be a Mover who himself is unmoved, an Unmoved Mover who begins the initial motion within the universe. That Unmoved Mover is God.

3. God the "Necessary Being"

Things in this world come into existence and eventually go out of existence. Continents come and go, plants and animals are born and die, buildings fall apart or are destroyed. Everything in this world is contingent. For each contingent being or thing there was a time when it did not exist and a time when it will cease to exist (in this world). However, not everything in this world can be contingent, due to the fact that given enough time, nothing would exist. Yet, things continue to exist everywhere in the world. Therefore, there must be someone who is not perishable, a Necessary Being behind the existence of all things that we see. That Necessary Being is God.

4. God the "Designer"

A large part of the world is filled with non-intelligent material objects that lack knowledge. These objects do not have brains. Yet, they behave in ways that demonstrate intelligence. Their behavior also cannot be due to pure chance since their activities follow very "predictable" paths. It appears as if they are being directed by someone with intelligence. This universal, unseen designer who directs the myriad of "non-living things" in nature is God.

5. God's Use of Math

The universe is not chaotic. The universe is orderly. That is why scientists are able to build a structure of understanding, putting one discovery within nature upon another.

The most incomprehensible thing about the universe is that it is comprehensible.

A. Einstein

One evidence of this kind of order and intelligibility in nature is that it follows mathematical principles. This idea is contained in a saying of Galileo:

> The book of nature is written
> in the language of mathematics.

Galileo in 1623

One evidence that math is embedded in nature is the fact that its equations and laws are discovered in different places by individuals unknown to each other. For example, the Pythagorean theorem was discovered in ancient Babylon about 1900BC (Plimpton tablet) and then later, independently by Pythagoras in Greece about 525BC. The math discipline of Calculus was discovered in the 1600s in Germany by Gottfried Leibniz and independently in England by Isaac Newton.

Mathematics by itself has predicted the existence of various things in nature *before* they were discovered by science! Four examples of this are the following.

1) The planet Neptune was discovered initially in 1846 by the separate math calculations of John Adams at Cambridge and Urbain Le Verrier in Paris, France.

2) Radio waves were predicted in the equations of James Maxwell of Scotland in 1862 and later discovered by Heinrich Hertz in Germany in 1887.

3) Quark particles within protons were discovered mathematically by Murray Gell-Mann at Cal Tech near Los Angeles and independently by George Zweig in 1964. Quarks were then proven to exist by using the Stanford Linear Accelerator in 1968.

4) The Higgs-Boson, or matter particle, was predicted mathematically by Peter Higgs in England and independently by Francois Englert in Belgium in 1964. The Higgs particle was, however, not discovered until 2011 by experiments using the Large Hadron Collider in Switzerland.

God and the mathematics he creates predate the world. God himself is not one being among many beings within the universe. God is not the highest of beings. God is not within the world he created. God is rather the very *Act of Being* itself, the fundamental reason or ground that makes something rather than nothing possible. God is the reason the universe is comprehensible.

6. God the "Programmer"

In 1953 James Watson and Francis Crick deciphered the structure of the DNA molecule which is present in the nucleus of all living cells. The DNA molecule looks like a tiny twisted ladder. The steps of the ladder consist of four chemicals (A, G, C, and T) called "bases." One DNA molecule has 3 billion of these "bases" in its structure! These "bases" are like alphabetic letters in a sentence. The arrangement of these "bases" carries "information" (called "genes") on how the cell is to build the body's basic protein structures. The living cell is, therefore, a complex information processing system.

Every DNA molecule is packed with "digital code" to do various things within the body. "Information" of this kind is known to always have been produced by some form of "intelligence." Examples of such information could be the writing on an ancient temple, or a paragraph in a modern book, the script of a radio broadcast, or the digital program within a computer. In all these examples, an intelligent being will have produced the "information." Information does not emanate from material things of and by themselves. In the example of the living cell (with its 3 billion bits of code) one sees an enormous complexity and planning. Such complexity points to the presence of an "omniscient" mind, or massive source of intelligence. That powerful mind is God.

CREATION

7. "Big Bang" Discovered ...
"In the beginning"- Gen 1:1

The "Big Bang" was discovered in 1927 by Father Georges Lemaitre, a Belgian priest. Father Lemaitre discovered in Einstein's equations that the universe with its many galaxies was expanding outward from a single point. What happened at this point of origin was later termed the "Big Bang." This mathematical discovery was confirmed in 1929 by Edwin Hubble while using the 100-inch telescope at Mount Wilson near Los Angeles. At the moment of the Big Bang, matter and space came into existence, and eventually many years later stars and galaxies would begin to form from this matter.

The universal expansion from the "Big Bang" can be pictured as a balloon being inflated with air, with red dots on the balloon representing the various galaxies. As the balloon is inflated, the red dots move further and further away from each other; meaning that, from any single point, everything expands outward. Then, if one were to run this movement backwards in time, by letting the air out, everything would return to the point of the Big Bang.

With that information in mind, one can visualize the existence of God. God is the only power capable of bringing matter into existence from nothing. God is the "First Cause" of all things coming into existence, or in existence - who makes everything that is not himself come into being.

What follows is an account of how the concept of an "expanding universe" was initially discovered and how it led directly to the scientific picture of the "Big Bang." Here are the steps.

(1) The French comet hunter, Charles Messier (1730-1817), discovered fixed cloud-like structures in the night sky. By 1781, Messier had catalogued 103 such "nebula." M-31 was the Andromeda nebula. He noted that nebula did not move across the night sky. These unmoving nebula remained a mystery into the late 1800s.

(2) In 1880, Henry Draper, an American physician, made the first "photograph" of a nebula. He used an 11-inch refracting telescope and camera to make a 51-minute exposure of the Orion Nebula. Such long exposures made very distant objects become visible on photographic plates. Very distant objects could then be studied for the first time.

(3) Henrietta Leavitt studied photographic plates of stars at Harvard University between 1907 and 1921. She identified 47 "Cepheid Variable" stars that are more massive than our sun and about 100,000 times brighter. Leavitt found that all Cepheids had the same absolute brightness. In other words, these stars have a consistent brightness no

matter where they are located. Some of these stars also exist within our galaxy and their relative distances were known. Thus, the amount by which a Cepheid star's brightness is dimmed by distance allows for that star's distance from the earth to be calculated. Leavitt published her findings in 1912. Edwin Hubble used her work in 1923 after he discovered a Cepheid star to be present within the Andromeda nebula.

(4) In 1912, Vesto Slipher at the Lowell telescope in Arizona discovered that nebulae are traveling at very high rates of speed. He studied 15 different nebulae by examining the red shift in their spectral lines. He learned that blue shifted nebulae were moving toward us while red shifted nebula were moving away. He was the first person to develop this method of determining the velocity of stellar objects. He published his findings in the Lowell Bulletin (#62) in 1912. He calculated that the Andromeda nebula was traveling at about 300 kilometers per second through space. In 1914, he further discovered that various spiral nebulae were turning in a circular motion on their own axis.

(5) In 1917, the 100-inch reflecting telescope was completed on Mt. Wilson outside of Los Angeles in California. Edwin Hubble began his work at Mt. Wilson in 1919. The Mt. Wilson telescope was the most powerful telescope in the world at that time.

(6) In 1923, E. Hubble was able to resolve individual stars in the Andromeda nebula. Furthermore, he found the presence of Cepheid Variable stars in the Andromeda cloud. Using the known absolute brightness of these stars, he was able to determine that the Andromeda stars were about 900,000 light years away from our Milky Way galaxy. This made Andromeda the first galaxy to be discovered outside of the Milky Way. Hubble went on to find that many other nebulae were also individual galaxies.

(7) In 1929, after studying the red shifted spectrums of 46 galaxies, Hubble announced the science of the "expanding universe." He stated that the galaxies farthest away from us were moving the fastest. This was expressed in an equation known as Hubble's Law. Hubble's work revealed a universe that is continually expanding away from a single point of origin. Though a telescope could not detect God's work within the initial seconds of the Big Bang event, the Bible has affirmed God's presence.

… the ages [or world] were put
in order by the Word of God …

Hebrews 11:3

the mystery hidden for ages
in God who created all things …

Ephesians 3:9

8. The "Big Bang" …
God as Creator

The Big Bang event was the beginning of the universe we live in. It came into being from nothing, in a set of orderly steps. Of this beginning, Arno Penzias, a Nobel Prize winner for his discovery of the Big Bang's background radiation, offered the following words.

Astronomy leads us to a unique event, a universe created out of nothing, one with the very delicate balance needed to provide exactly the conditions required to permit life (to exist) and one which has an underlying plan.

A. Penzias, recorded by W. Bradley
in *The Designed "Just So" Universe*, 1999.

The Big Bang occurred about 13.8 billion years ago. It is estimated that the initial temperature of this event was roughly 4 trillion degrees Celsius (about 250,000 times hotter than the center of our sun).

The first seconds of the Big Bang saw a quark-gluon plasma come into existence. In the center of all atoms are protons and neutrons. Each proton and neutron is made of three quark particles glued together by "massless" gluon particles. Gluons have existence but no mass. They are an early sign of human life beyond death.

The second step within the Big Bang occurred between 5 to 20 minutes after the initial start. It involved the bringing together of protons and neutrons to form the "nuclei," or the centers of atoms. All "atoms" consist of a nucleus, plus electron particles flying in a cloud-like arrangement around the atom's center. However, the early Big Bang event was too hot for electrons to be positioned around the atom's center. Electrons were present but only as part of the very hot plasma soup.

The third step of the Big Bang event began after the early "nucleic soup" had cooled significantly. This took about 380,000 years! Then, things were cool enough for whole atoms to form. The first hydrogen and helium "atoms" came fully into existence. That is, the very tiny "electron" particles began to fly around the centers of these "atoms." And thus, with the formation of whole atoms, stars and planets began to form.

The hydrogen atom is the simplest atom in nature. It is composed of one proton particle in its center and one very tiny electron particle flying around it in a cloud-like fashion. It was the most abundant atom in the early universe, and even today makes up about 90% of all atoms in the universe.

The hydrogen "nucleus" (of atom #1) had one proton and no neutrons. However, another form of the hydrogen nucleus, called "Deuterium," had one proton together with one neutron in its makeup. Two of these Deuterium nuclei then fused together to form the Helium nucleus (of atom #2). This was followed by the Lithium nucleus (of atom #3) with its three protons. These three kinds of nuclei were the ingredients of the dense, soup-like mass that took so long to cool off.

All the hydrogen "atoms" in the universe of today were essentially formed in the Big Bang event. Today, when we drink a glass of water (H20) we are imbibing hydrogen atoms made in the Big Bang. Since the human body is about two-thirds water, it is evident that we are carrying around part of the Big Bang in our bodies. In other words, a significant part of our body was made 13.8 billion years ago! With this in mind, no one should say that God did not have a plan, or that he has not cared for us. He has been there all along the way!

Another bit of evidence of God's presence in the Big Bang development comes under the heading of "fine tuning." Fine tuning refers to the very tight parameters present in the formation of the universe. There are about 30 different kinds of fine tuning necessary for living things to be able to exist on this planet.

One example of fine tuning is the "strong nuclear force" which holds positively (+) charged protons together in the nucleus of atoms (even though the protons strongly repel each other). The strength or level of this "strong force" has to be very precise or the universe would be a very different place than it presently is. If the "strong force" was just a little *stronger,* two protons would not separate. The "strong force" would permanently fuse two protons together; thus, making the element "helium" in the process. If the "strong force" had fused all the hydrogen into helium nuclei (element #2) in the Big Bang, then there would be no hydrogen today. This would mean that there would be no water (H20) to make living organisms possible.

If, however, the "strong force" was a little *weaker,* then the nucleus of atoms would not stay together! In this case, the repulsion of the positively (+) charged protons would destabilize the whole nucleus. The nucleus would then tend to fly apart. In such a case, the only stable element in the universe would be the hydrogen atom itself, because it has only one proton. The other atoms that support life, such as carbon, oxygen, calcium, phosphorus, nitrogen, and iron, with their many protons, would be unable to form. The result would be that we would not be here!

Thus, one sees how God has kept his "hand" on the throttle, guiding his universe very precisely!

9. The Formation of Stars ...
Light and Warmth for Us

Hydrogen is the main fuel in stars. Stars produce energy by changing the element hydrogen into the element helium. In this reaction, some of the mass of the hydrogen atom (about .7%) is changed into energy.

Hydrogen → Helium +   energy (7/10ths of 1%)
(element #1)   (element #2)   as light & heat

The light and heat from our star, the Sun, comes from the energy produced in the above reaction.

Stars begin their life by the "inward pull" of hydrogen atoms due to the force of gravity. The force of gravity causes huge pressures to be built up in the centers of stars. These pressures cause the star to get hotter and hotter until at around 5 million degrees Kelvin (roughly 9 million degrees Fahrenheit) its hydrogen nuclei begin to fuse inside the star's core. At this point, the new star begins to "shine," giving off heat and light.

Jesus said at one point in his ministry that he was the "light of the world" (John 9:5). Science has found that the "light" particle is the "photon." The photon is a "mass-less" particle, generated from the movement of "electrons.". The "photon" exists within the atom but is without mass. Jesus, like light, is everywhere present. He is the author of stars and of light. He has gifted us with the light and warmth needed for us to live on this planet.

The heavens are proclaiming the glory
of God and the works of his hand …

                                              Psalms 19:1

God appoints the number of stars.
He calls to them all by names.

Psalms 147:4

"I [as a scientist] was merely thinking God's thoughts after Him. We astronomers are priests of the Highest God in regard to the book of nature."

Johannes Kepler

10. The Formation of the Elements ...
Atoms for the World

Stars and galaxies began to appear in space about half a million years after the Big Bang event. After the Big Bang, about 90% of the matter in the universe was element #1, hydrogen. Element #2, helium, made up only about 8% of all matter. A small amount of lithium, element #3, was also present. The other 89 basic elements making up the world around us today were yet to be made within the newly forming stars.

Stars are "element factories," the places where complex atoms are made. As the hydrogen gas collected into huge cloud-like balls under the influence of gravity, their inner core temperatures reached extreme levels. At some point, hydrogen nuclei were moving so fast that they slammed together with sufficient force to form helium nuclei. Then, after a star used up all its hydrogen fuel, it began to fuse its helium fuel. Helium nuclei were transformed thereafter into heavier forms of matter. For example, three helium (4) atoms fused together to form one carbon (12) atom. And four helium (4) atoms fused together to form one oxygen (16) atom. The events looked something like this ...

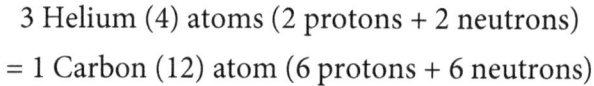

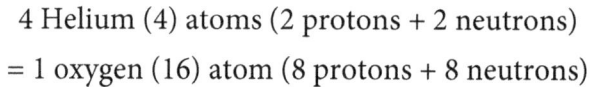

3 Helium (4) atoms (2 protons + 2 neutrons)
= 1 Carbon (12) atom (6 protons + 6 neutrons)

4 Helium (4) atoms (2 protons + 2 neutrons)
= 1 oxygen (16) atom (8 protons + 8 neutrons)

When a star finished fusing helium, it would fuse other, heavier nuclei, forming shells surrounding the core of the star. One example of this is the event of two carbon [12] nuclei forming one magnesium [24] nuclei.

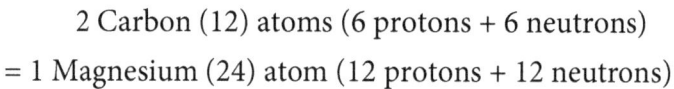

2 Carbon (12) atoms (6 protons + 6 neutrons)
= 1 Magnesium (24) atom (12 protons + 12 neutrons)

When the core of a star contained mostly iron (element #26), the fusion process stopped. Elements above #26 were then made increasingly larger by a process called "neutron capture." Neutrons have no electrical charge (n) and therefore can attach themselves to other elements and thereby increase that element's overall size. For example, a neutron can change into a proton particle by ejecting one electron to make itself positively charged and then join another element to make it larger.

The neutron capture process involves all of the elements from 27 to 92. The "number" of a specific element reveals the number of protons in its nucleus. For example, the iron atom [element #26] has 26 protons in its nucleus.

In giant stars that explode as supernovas, the process of making heavy elements is very rapid. For example, a typical supernova explosion may last only a few weeks, yet during that time all the heavier elements will have been created and blown outward into space. Our Sun, that was formed 9 billion years "after" the Big Bang event, is one of the newer stars that has used these heavier elements in its formation. Our Sun, and the planets that surround it, have acquired many of their heavy elements from earlier supernova explosions and the "neutron capture" process in smaller stars.

The atoms that are made inside of stars are also the building blocks of the human body. The human body contains about 25 of the 92 basic elements in nature. Each of these 25 different atoms performs specific functions in our bodies. The following six elements make up 98% of the human body by weight.

oxygen (#8)   In "respiration" to release energy for the body.

carbon (#6)   Primary component in many body structures.

hydrogen (#1)   In water for blood and bodily fluids.

nitrogen (#7)   In amino acids to make proteins, especially DNA.

calcium (#20)   In bones and teeth.

phosphorus (#15) In bones and ATP, the energy molecule within all cells.

Another five atoms make up about 1.5% of human body weight and are required for life.

potassium (#19)   For cell function and heartbeat.

sulfur (#16)   In amino acids, hormones, enzymes, and antibodies.

sodium (#11)   Regulates water balance in the body. In nerve and muscle reactions.

chlorine (#17)   In bodily secretions.

magnesium (#12)   In 300 enzymes, and in all body cells.

The final list of atoms in the human body are called "trace elements." They make up about one-half of one percent of the body by weight, while doing important work.

cobalt (#27)   In Vitamin B12. Essential in hemoglobin formation.

chromium (#24)   Role in insulin usage.

copper (#29)   Crucial for forming red blood cells.

	Kills bacteria and viruses.
fluorine (#9)	Helps prevent cavities in teeth.
iodine (#53)	Important in thyroid hormones.
iron (#26)	In hemoglobin. Carries oxygen from lungs to all cells.
manganese (#25)	In connective tissue. Also, a blood-clotting factor.
molybdenum (#42)	In amino acid metabolism.

One final point. Every atom in the universe presents itself in just three ways: as a solid, as a liquid, or as a gas. For example: An iron atom is a solid at 70 degrees F., a liquid at 2800 degrees F., and a gas at 5200 degrees F. This 3-in-1 structure is an image of the Creator and can be clearly seen in the Bible. In the Bible, the one God is revealed in three subjects: Father, the Father's Word (Jesus), and the love between them, the Holy Spirit. Furthermore, this three-ness is also mirrored in humans: one person, his or her own word, and the great love between the person and his or her word. (Notice, for example, how we sometimes "cut other people off" to utter our beloved word.) Furthermore, this word can also go out from us into the world, as Jesus went out from the Father, to do his work in the world.

Like a great artist, whose own image is seen in his work, God's 3-in-1-ness is imaged both within his creation at large and in his creatures.

If you had known me [Jesus],

You will also have known my Father

... I and the Father are One

John 14:7 & 10:30

I [Jesus] am in the Father and the Father is in me ...

John 14:11

The Holy Spirit, whom the Father will send in my name, will teach you all and will remind you of all that I [Jesus] said to you.

John 14:26

11. The Discovery of the Atom ...
Invisibility of Matter

In the year 1900, the atom was thought to be the smallest solid particle of matter, and, like the universe, had always existed. These two beliefs were incorrect! By 1950, scientists knew that the universe had a beginning in the Big Bang and that atoms were composed of a family of particles contained in mostly empty space. What follows is a brief history of the discoveries that led to our present understanding of the atom.

Date	Discovery	Discoverer	Location
1897	Electron particle discovered	J.J. Thomson	England
1909	Nucleus of the atom found (Gold leaf experiment)	Rutherford (New Zealand)	England
1913	Atom's structure is visualized. A nucleus with electrons flying around it	Rutherford & N. Bohr	England Denmark
1919	Proton particle is found in the nucleus	Rutherford	England
1932	Neutron particle is found in the nucleus	J. Chadwick	England
1938	How stars shine by the fusion of hydrogen nuclei	Hans Bethe (Germany)	U.S.A.
1957	Chemical elements are formed inside of stars	Fred Hoyle & W. Fowler	England
1964	Quark particles predicted to be inside protons and neutrons	Gell-Mann	U.S.A.
1968	Quark particles found	Stanford Linear Accelerator	U.S.A.
1976	Gluon particles predicted	Mary Gaillard & Graham Ross	U.S.A.
1979	Gluon particles found	Tasso group	Germany

12. The Origins of the Solar System
The Earth as our place to live

The Solar System of which the Earth is a part began to form about 9 billion years *after* the Big Bang event. Many populations of stars had come and gone in that first 9 billion years. Those stars had begun by fusing hydrogen atoms together in their very hot cores. Then, as time went by, they fused heavier atoms into still heavier atoms until they had exhausted all their various fuels. Finally, at the end of their life cycle, they blew off into space these newly formed, heavier elements (or atoms as we call them).

Later, when the Solar System began to form, these heavier elements were gathered together as "dust," together with the basic hydrogen gas, and began to form into a large cloud-like structure under the influence of gravity. As this massive cloud-like structure began to pull in upon itself, it became hotter and hotter, and gradually began to spin. The center of this mass [99%] would eventually become the Sun, and the little swirls of material that spun off from the Sun would eventually coalesce further and become the planets, including our Earth. This all began to look like our Solar System about 4.5 billion years ago.

The early Earth was very hot and molten for a long time. Not much is known about this period. However, about 4.4 billion (or 4,400 million years ago), the earth had cooled and solidified enough so that landforms could begin appearing. At first, these landforms were volcanic islands, like those making up the Hawaiian chain of islands today.

These volcanic landforms rode upon the still hot mantle material of the earth's crust. Eventually, these volcanic islands would come together to form the first continent upon the earth's surface.

This whole story of the earth's early formation demonstrates the continuing presence of God. The more obvious evidence of his work is sometimes referred to as "Fine Tuning." What follows are a few examples of how God has "fine-tuned" the earth for its use by humans.

Gravity within the Sun

In nature "gravity" is a relatively weak force, as compared to the Strong Nuclear Force that holds an atom together. Gravity, however, is strong enough to cause the gradual collapse of hydrogen gas and dust into a highly compressed, very hot ball of fusible matter. And from this gravitational beginning, stars, like our Sun, would begin to form and eventually shine.

The force of gravity, however, must lie within a very narrow range. If the force of gravity was slightly *higher* in strength, or slightly *lower*, the universe that we live in would not have come into existence. For example, if gravity was *greater* than it now is, the center of stars would be more highly compressed and therefore much hotter. Stars would burn (or fuse) their hydrogen fuel much too fast to last very long. Such a star might go through its life cycle in only a million or a few million years. This period of time would not have been long enough to allow for the development of intelligent life like ourselves. The Earth, for example, is about 4.5 billion years old, yet humans have been on this planet for only about 2 million years.

On the other hand, if the force of gravity had been somewhat *weaker*, stars such as our Sun might never have formed at all. Units of hydrogen gas and dust might never have pulled together enough to start nuclear fusion. Therefore, many of the heavier elements, necessary for life, would not have been formed. The result would have been that we would not be here!

Gravity on Earth

The relative weakness of the gravitational force is crucial for life on the Earth. If gravity was *stronger*, bodies the size of today's humans would not be possible. Bodies our size would have been crushed by gravity. In a system with strong gravity, even insects would need thick legs to support their weight. Animals, in such a system, would in fact be very small in size.

On the other hand, if gravity had been somewhat *weaker* than it now is, creatures like us could not stay upon

the Earth's surface. We would be flung off into space because the earth is turning (at the equator) about 1000 miles
per hour. By comparison, gravity on the Moon is only one-
sixth that of what it is on earth, so the astronauts on the
moon in 1969 could jump high off its surface with very little
effort.

Earth's Distance from the Sun

The Earth's position with respect to the Sun is not too
close to be burning hot or too far away to be freezing cold.
The inner planets, Mercury and Venus, are too hot for most
living things to exist there. However, the outer planets beyond Mars are too cold for our kind of life to exist there.
The Earth, therefore, is in what is called "the sweet zone,"
or what some call "the Goldilocks zone." We get plentiful
warmth from the Sun, but not too much. And we are not
too far away so that our surfaces are entirely frozen. We are
situated in the life zone.

He is the God not of the dead, but of the living.

                                          Matthew 22:32

Temperatures in the Solar System:

Inner planets: Mercury +801 degrees F. and Venus +864
                degrees F.

Outer planets: Europa, a moon of Jupiter: minus 260 degrees F.

Titan, a moon of Saturn: minus 290 degrees F.

Earth's Magnetic Shield

The Earth is like a giant magnet with both north and south "poles." The Earth is surrounded by magnetic lines of force which keep the Sun's "solar wind" from hitting us unimpeded.

The Sun's solar wind is composed mostly of proton particles traveling at *great* speed. In other words, it consists of a continuous stream of high energy particles. Its presence is occasionally evidenced on Earth in the "Aurora Borealis," and in its occasional interference with radio and TV broadcasting. It is also seen in comets as they pass by the Sun with their tails always pointing away from the Sun's outer surface. (The planet Mercury has had most of its atmosphere stripped away by the solar wind.) Without this continuous protection of our magnetic shield here on Earth, humans could not live on this planet.

By studying how shock waves from earthquakes travel through the earth, the inner structure of the Earth has been detailed by science. The center of the Earth consists of a solid iron core about two-thirds the size of the Moon. Surrounding this core, at lower pressure and temperature, is a liquid iron and nickel slurry that moves about as the Earth spins around on its axis. This flowing liquid metal generates electrical effects which in turn produces the magnetic field that is seen around the Earth. It is this magnetic field that forces the Solar Wind to flow around the Earth, rather than impact it directly.

Once again one sees the planning hand of God as he saves us from a potentially dangerous situation.

The Moon Stabilizes the Earth's Climate

The Earth-Moon system is really a double-planetary system, thought by many scientists to be a very rare arrangement. The Moon helps to stabilize the rotation of the Earth.

When the Earth rotates, it wobbles slightly back and forth on its axis, like a top. The Earth rotates on its axis, tilted at an angle of 23 1/2 degrees relative to the Sun. Without the Moon, this axial "tilt would vary" dangerously over time. Presently, it varies only about 1%, but without the Moon's stabilizing influence this variance would be 10-15%. The seasons would thereby be seriously affected. Summers, when we lean toward the Sun, would be longer and hotter. Winters, when we lean 23 1/2 degrees away from the Sun, would be colder and longer.

The Moon's presence also lightens up our nights here on earth. It is a friendly companion.

The invisible things of him from the creation of the world are clearly seen, being understood by the things he [God] has made …

Romans 1:20

BIOGENESIS

13. The Early Earth ...
Water for Our Life

The Solar System was formed about 4.5 billion years ago. The Sun comprises about 99% of the total mass of the Solar System. The sun is about 110 times wider than the earth. The other 1% of the material in our solar system became the various planets that surround the Sun, one of which is the Earth upon which we reside.

About 100 million years after the formation of the Earth (4.4 billion years ago) the planet that we live upon had cooled enough for liquid water to exist on its surface. This is known from the discovery of water-bearing rocks in western Australia. Zircon is a rock that can only be formed in the presence of liquid water. Zircons discovered at Pilbara in western Australia date to 4.4 billion years before our time.

The existence of water is crucial for the development of living things. The water molecule has two electron openings in its outer structure. It seeks to share these two openings with other compounds so as to fill its outer shell. This means that it is a solvent, able to combine with many different chemicals. That is especially helpful for us because our bodies have many diverse chemical needs. In fact, water makes up about two-thirds of the human body by weight.

Most of the Earth's surface is also covered by water. About two-thirds of the Earth's surface is immersed in water that is on average about two miles deep. The salinity of the ocean is about 3%. The salinity of our bodily fluids is about 1%.

Where did all this water on our planet come from? Science is not sure, but here are some possibilities.

(1) About 90% of the atoms in the Universe are hydrogen, formed in the early Big Bang event. The oxygen atom was formed later, within the many generations of stars, and before the Sun's appearance. (4 Helium atoms = 1 oxygen atom). When the early Earth had cooled below 212 degrees F. (the boiling point of water), liquid water could exist; then hydrogen and oxygen atoms may have simply joined up to form the H2O we call common water.

(2) Another theory about the origins of water is that certain water-rich comets and asteroids impacted the early Earth and deposited large amounts of water here. The many craters on the Moon, for example, evidence these early impacts, especially from asteroids. Such evidence, however, is not easily found here on earth due to water erosion, over long periods of time.

[3] A third possibility for the large amounts of water found on Earth is that it might be the result of volcanic activity. Hot liquid "magma" from within the Earth contains significant amounts of water. Furthermore, the early Earth

had many volcanic islands, like Hawaii, that probably sent large amounts of water vapor into the air. This water vapor would then have become rain and would have filled the ocean basins.

One final thought. Notice in the NT how Jesus walked upon the water of the Sea of Galilee (Matthew 14:26) and changed 120 gallons of water into wine at the marriage feast in Cana (John 2:9). God has power over the water molecule. Since our body has considerable water in its structure, we should turn to Jesus when we too need help.

14. History of the Continents …
and the Arrival of Oxygen

As the earth cooled, a crust formed over the earth. This crust is like the skin of an apple. It is very thin: about 5 miles thick under the oceans and about 25 miles thick under the continents. Below the crust is the "mantle." Cracks in the mantle bring up hot "magma," or liquid rock, from within the earth. These openings in the mantle can be seen in the eruption of volcanoes.

The Earth's crust is not one solid piece. Rather, the earth's crust is cracked into a number of pieces, like the pieces of a jigsaw puzzle. These pieces are referred to as "plates." There are seven major plates that make up 94% of the Earth's surface. In total there are twelve plates which move about on the Earth's crust. What "moves them" is volcanic material which appears between the plates that are moving apart. As the volcanic magma comes up from the mantle below, the plates separate. This separation causes plates on the opposite side to run into each other, one going up to make mountains and the other going down to cause earthquakes. These complicated movements are called "plate tectonics."

Plate tectonics continually enriches our world with new minerals from the interior of the Earth. Plate tectonics represents a kind of recycling of crustal materials. The plates migrate around the Earth's surface forming new arrangements while bumping into one another. The plates migrate about as fast as one's fingernails grow, an inch or so each year. Therefore, not much movement will be seen in a single person's lifetime. However, over long periods of time these movements manifest themselves.

The history of the continents began as volcanic islands, not unlike the Hawaiian Island chain of our day. As time passed, these island chains, riding on the backs of crustal plates, came together to form **the Earth's first continent, "Vaalbara."** Vaalbara formed about 3.6 billion years ago and lasted until about 2.8 billion years ago. Then the moving plates broke this continent apart into separate, smaller continental fragments that do not resemble anything we see on today's maps.

During the time of Vaalbara, the first life forms appeared on Earth. These were the "cyanobacteria," which form coral-like "stromatolite" formations on the ocean floor. They first appeared about 3.5 billion years ago and can still be seen alive today in Shark Bay, Australia. These prokaryote bacteria were the first living things on earth to do photosynthesis, that is, to make free oxygen, so that later forms of life on Earth would have enough oxygen to breath. Cyanobacteria are often referred to as "blue-green algae." This is an error. They are not algae. They are bacteria which appear blue-green (cyan) in color. Photosynthesis, that appeared with these bacteria, works in the following way.

$$6CO_2 + 6H_2O \xrightarrow{\text{Sunlight}} C_6H_{12}O_6 + 6O_2$$

Carbon Dioxide   Water   Glucose (Energy)   Oxygen

Cyanobacteria use the sugar they produce as food. For them, the oxygen they produce is a waste product. Cyanobacteria, and various simple plants, produced oxygen for nearly three billion years, readying the atmosphere for the arrival of large animals.

"Respiration," or energy production in animals, is the opposite of photosynthesis. Respiration in humans resembles the following.

Glucose + oxygen → yields energy + CO2
+ Water (in our urine)

(Sugar our basic   [to exhale
    fuel)   carbon]

Vaalbara broke up about 2.8 billion years ago. Then, for about a 100 million years, the land masses on Earth existed separately. These landforms did not look anything like the continents of today; their shapes were very different.

The crustal plates kept on moving, while carrying the separated land masses on their back. Then, eventually, about 2.7 billion years ago, **a second supercontinent** was formed. This has been given the name **"Kenorland."** It included all or most of the land areas on earth. It lasted until about 2.1 billion years ago.

During this time the first complex, multicellular plant-like forms appeared on Earth. Earlier the cyanobacteria had been very small, did not have a nucleus, and reproduced by simply dividing in two. They were (and still are) known as prokaryote cells. Then, about two billion years ago, the first eukaryote forms appeared on the Earth. One of these plant-like forms has been found in fossils in an iron formation near Marquette, Michigan. It is the algae "Grypania spiralis." Its fossils are flat discs 5 inches across and up to 18 inches long. Hundreds of specimens have been found. Eukaryote organisms have a cell nucleus with DNA, make their own food and produce oxygen through photosynthesis.

A third supercontinent, "Columbia," formed about 1.8 billion years ago. It lasted until about 1.4 billion years ago. During this period the free oxygen level in the atmosphere grew to about 10% (today it is roughly 21%). This made it possible for multi-celled organisms and eventually larger animals to live on this planet.

Also, during this time, the ozone layer began to develop overhead. Ozone is a molecule of 3 oxygen atoms hooked together (O3). It forms about 10 to 20 miles above the Earth's surface. The ozone layer shields living things on Earth from the harmful ultraviolet B radiation of the Sun. In humans this short-wave radiation can cause skin cancer and cataract formation in the eyes. Ozone is another helpful part of God's plan to protect the development of life on this planet, and especially to protect humans.

A fourth supercontinent, "Rodinia," developed about 1.1 billion years ago, and lasted until about 750 million years ago. During this time small multi-celled water-based organisms became widespread in the Earth's oceans. Also, sexual reproduction became established, so that genes from two parents could be combined in the next generation. This mixing of genes speeded up changes in the structure of organisms within the oceans. It also protected living things from disease-causing agents. The landforms of Rodinia, however, were still barren of life. Life had not yet colonized the dry land of the planet.

A fifth supercontinent, "Pannotia," centered on the South Pole, and formed between 650 and 550 million years ago. In the later years of Pannotia the first of the larger animals, including shelled invertebrates (animals without a backbone) and soft-bodied creatures like jellyfish, appeared within the ocean. It was the beginning of an explosion of lifeforms on Earth, the so called "Cambrian Explosion."

A sixth semi-supercontinent, "Gondwana." appeared between 550 and 335 million years ago. It consisted of three large continental areas in close proximity. On this landmass the first fish, amphibians and reptiles appeared.

A seventh supercontinent, "Pangaea" (or "Pangea"), appeared about 335 million years ago, and began breaking up about 175 million years ago – and is still doing so today! This supercontinent included today's continents of North and South America, Europe, Asia, Africa, Australia, and Antarctica. It broke up in two steps: (1) About 175 million years ago, North America and Europe separated. The North Atlantic Ocean was formed as they separated. However, the future South America, Africa, Australia, and Antarctica remained together in one land mass. (2) Then, about 140 million years ago South America and Africa began separating from each other. The South Atlantic Ocean was formed by this separation. Also, about this same time, India began migrating away from East Africa toward Asia, crashing into Asia about 35 million years ago, and forming the Himalaya

Mountains. Australia also broke away from Antarctica and is presently moving northward.

Evidence for such continental movements is two-fold: (1) The same fossil plants and animals are found on continents now separated by thousands of miles. For example, the fossils of Cynognathus, a mammal-like reptile of 240 million years ago, have been found in both South America and Africa. Also, the fossils of "Lystrosausus," another 250-million-year-old ancestor of mammals, have been found in South Africa, India, and Antarctica. (2) In addition, on a world map, the matching coastlines of Eastern North America and Europe and the close fit between South America and Africa can be seen. The soil types and landform features of these separated continents also show a near identical congruence.

Though people today have a better scientific idea of how and why the continents are moving, the planning and power of God is *still* evident in the earth's history.

God's hand has done this. In whose hand is …
the breath of all the life of man.

Job 12:9-10

15. History of Plants ...
Food for Humans

Green plants manufacture their own food. They take water (H_2O) from the soil and carbon dioxide (CO_2) from the air, plus energy from the Sun, and combine these ingredients to make basic sugar ($C_6H_{12}O_6$). They do this by the process of photosynthesis. Sugar therefore is the basic plant food. That is also why most vegetables and fruits taste sweet. All carbohydrates, such as wheat and rice, are composed of long chains of these sugar molecules. This also explains why humans are attracted to sweet food products. Land plants have a half-billion-year history. All of the plants grown today by farmers around the world are connected to that history.

The history of land plants began modestly with the appearance of brown algae, or "seaweed," on rocky shores. This was followed by one of the first true "on-the-land" plants – namely, the "mosses." Mosses are small, flowerless plants, composed of simple leaves often one cell thick, attached to a short stem. They grow in mats in wet areas: bogs, on the banks of streams, and in moist forests. After fertilization, they produce "spores." Fossil mosses have been found and dated to about 450 million years ago.

The second advance in the history of plants was the appearance of vascular tissue, in such plants as ferns. One of the earliest vascular, or stem plants, was "Aglaophyton," which appeared about 425 million years ago. Vascular plants possessed two advances: (1) They produce "lignin" that makes their cell walls rigid and supportive. This feature allowed plants to grow much larger. (2) They also have "xylem" tubes that transport water and minerals from the soil, and "phloem" tubes that transport sugars throughout the plant body. Ferns, like mosses, reproduce by means of spores that require the plant to remain close to water sources.

The third advance in the history of plants was the appearance of "trees" with woody stems. Only after vascular tissue and roots were in place could trees appear. The first trees have been named "Wattieza" and their fossil remains date to about 385 million years ago. They stood about 25 feet tall and had fernlike leaves. They also were part of the world's first known fossil forest, located at Gilboa, in New York state. Like mosses and ferns, they too reproduced by spores and therefore had to be located near water sources (sperm require water to swim toward an egg cell).

The fourth advance in the history of land plants was the appearance of "seed." Seeds appeared about 350 million years ago. Seeds contained an embryo along with moisture and nutrients. Seeds allowed plants to colonize dry lands, rather than only wet areas. They also allowed the embryo to be transported away from its origin.

The first "naked seed" plants (plants with unprotected seeds) were the gymnosperms. Gymnosperm seeds lie unfertilized, in an open-air part of the plant, waiting to be pollinated. The "Cordaites" were the first gymnosperms, appearing about 325 million years ago. They have been found in the coal mines of Belgium and Germany. Then, about 280 million years ago another gymnosperm family, the Conifers, appeared. The seeds of Conifers lay naked on the scales of their cones. The Conifers included the Douglas Firs, Spruces, Redwoods, Cypresses, Junipers, and the Yews. (Pines would appear later).

The fifth advance in the history of land plants was the appearance of flowering plants. Flowers began appearing on this planet about 140 million years ago. A fossil imprint of the flowering plant "Ficus specrosissima" has been dated to 140 million years ago. In flowering plants, the flower surrounds the male "stamen," which contains pollen (male sex cells). A flower also surrounds the female "pistil" which consists of a tube with a round base, the "ovary." The ovary contains the female sex cell, or egg. When the egg is fertilized by male pollen, the ovary produces seed. This seed is "protected' inside the ovary. The ovary also produces food for the seed to use when it is ready to grow on its own. The ovary with its seed and food is called a "fruit." Humans eat many kinds of fruit. For example, apples, tomatoes, and avocados are fruits.

The pollen of the male flower has to be transported to the female ovary by wind, insects, or birds. When a plant puts out its flowers, bees often come to do their charitable work! Bees are so good at carrying pollen to female plants that about one-third of all food crops eaten by people on earth are fertilized by this insect. Bees appeared in the fossil record a few million years after the appearance of flowering plants.

Pine trees appeared in the fossil record about 153 million years ago. Pine trees are evergreen Conifer trees, having naked or unprotected seeds. Pine trees are found in almost all parts of the planet (in six of the seven continents). They have bluish-green leaves in the form of needles and grow to 130 feet. They are the leading source of paper and building materials for houses on this planet. One Conifer, the "Bristle Cone Pine," is the longest living plant on Earth: there is one such pine that is 4,700 years old near Bishop, California.

The "grass family" includes such food crops as wheat, rice, barley, and corn (maize). The first grass plants appeared on this planet between 65 and 80 million years ago. Fossilized grass pollen has been dated to between 65-70 million. It has been determined genetically that wheat and rice had a common ancestor about 50 million years ago. However, wheat and barley date only to about 4 million years before our time. Einkorn wheat, the earliest modern kind of wheat, has been grown in the Tigris-Euphrates river basin for between 10,000-40,000 years.

Root crops have also been popular for at least 5,000 years. Even before our ancestors learned how to plant seed to grow their own food, they foraged for wild forms of carrots, beets, turnips, and radishes.

One final point: scientists have determined that "roses" began to brighten up this planet about 37 million years ago. A flower show for the ages!

Consider the lilies of the field, how they grow and labor not.
I [Jesus] say to you, that even [King] Solomon in
all his splendor was not dressed like one of these.
But if God [the Father] so clothes the grass of the field …
shall he not much more clothe you, ones of little faith?

Matthew 6:28-30

16. History of Animals …
a Body for Humans

The Universe, at its conception, could not have "evolved" because there would have been nothing for it to have evolved from. The principle is this: From nothing, nothing will come. Therefore, God first created the world from nothing, then he let it evolve into what we see today, guiding it through the use of secondary causes. God would use nature's ingredients to bring about evolutionary changes.

Evolution is based on "natural selection," or selection by nature. This essentially means "survival of the fittest." The fittest are those plants or animals that live long enough to breed and pass on their genes into the next generation. Thus, each generation of a particular species will be better adapted to its environment than the generation that preceded it. Therefore, over a long period of time, each species will "evolve," or improve.

The evolution of living things can, however, be viewed in another way. Evolution can be based upon the notion that a rise in complexity comes from above. For example, (1) water and minerals in the soil are "raised up" by the plants that absorb and use them to grow. Plants, however, have no highly evolved sense organs or brain, and generally do not move about. Plants however, do "behave intelligently" because they are guided by God. (2) Plants are then, in time, eaten by animals, who do develop sense organs and a brain, and can move about. Animals also have the beginnings of intelligence and possess an "instinctive life" that is also guided by God. (3) Finally, humans find themselves placed above these earlier groups, taking into themselves both plants and lower animals, and thus raising them up to an even higher form of life. Humans possess an advanced intellect and free will. However, in spite of these advantages, human life on this planet is limited - being subject to death. God, however, will bring the human family a solution to this weighty problem. God will come down from Heaven to "raise us up," offering his people a new divine-like nature and everlasting life in Heaven (John 1:14; 6:33-35). This kind of Life is acquired through Baptism and the Holy Eucharist (the ultimate kind of Food ... namely, the Lord's own risen Presence). Humans can, therefore, live eternally in Heaven by accepting God's gift of his own Life within them (John 3:5; 1 Corinthians 10:16-17 & 11:29).

The continual improvement of various animal species, from one generation to the next, is also the foundation over time for the development of the human body. In other words, *through the history of animal life on this planet, people have received a physical body.* Each organ of the human person has its own specific history. The brain, eye, ear, hand, leg, heart, lung, digestive tract, kidney, and liver separately have their own specific history. The history of each body part can be traced within the animal kingdom – piece by piece.

The width of a human hair is about ten thousand atoms. And, amazingly, in the words of Jesus: "The hairs of your head have all been numbered" (Matthew 10:30 & Luke 12:7). Notice also that the "soul" has not been mentioned here. The soul forms the body and is created directly by God at the time of a person's own conception. In other words, each of us is specifically created one of a kind by God, unique in all of history.

Below is a chart listing **the origin of various aspects of animal life on this planet.** The dates given are based upon current knowledge of the fossil record.

Millions of years ago	
760	Sea sponge: no brain, nerves, circulation system or digestive tract
570	Shelled invertebrates (no backbone) …Trilobites very numerous.
518	First fish: "Meta spriggina" Gills, notochord, eyes on a wormlike body
425	First jawed fish: sharks
405	First insects: "Rhyniognatha hirsti"
400	Lobe-finned fish: "Coelacanth" Crawled up onto land. Still exist today.
385	Lungfish: "Dipterus valenciennes." Breathed air and crawled onto land. Living example: the Queensland lungfish
368	First amphibian: "Elginerpeton" Four limbs.
365	Amphibian: "Ichthyostega" from Greenland Lived at the edge of water and land.
350	Winged insects: flight begins on Earth
312	First eggs covered by a shell: "Protoclepsydrops. Made reproduction possible on land away from water.
310	First lizard-like reptile: "Hylonomus"

275	First dinosaur-like reptile: "Dimetrodon"
260	First mammal-like reptiles: "Lycaenops" Size of a wolf. Legs fully under body. Long slender head. Dog-like fangs. Walked/ran.
231	First small dinosaurs on land: "Eoraptor"
205	First Mammals: "Moganucodon." Shrew-like. Four inches long.
180	Flying reptiles: "Pterosaurs"
150	First birds, with teeth: "Archaeopteryx"
148	First placental mammals: "Eutherians" Baby is carried inside mother. Live birth.
116	Birds with beaks (no teeth)
110	First ducks: "Gansus yamenensis" Webbed-feet
100	First ants and bees
65	First primates: "Lemurs"
60	Modern bird lines
35	First monkeys: "Aegyptopithecus"
34	First elephants with trunk: "Palaeomastodon" Ancestor of all elephants.
27	Ape-like monkey: "Proconsul"
25	Raccoons
15	First horse: "Merychippus" Long legs and face. Ancestor of all breeds.
12	First wild cats
6	Upright walking apes
5	Wild boar: Ancestor of pigs
3	Wild sheep: "Mouflon"

2	"Aurochs": Ancestor of cattle
2	First gray wolves: Ancestor of dogs
1 (?)	Chicken: jungle fowl of China and India

17. History of The Organs of the Human Body

History of the Heart

Before I formed you in the womb, I knew you …

Jeremiah 1:5

Millions of years ago	A human heart weighs about one pound and pumps 2000 gallons of blood each day.
760	Sea sponge: no brain, nerves, circulation system or digestive tract
550	Amphioxus 1 chamber pump
425	Fish 2 chambers
310	Reptiles 3 chambers
205	Mammals (people) 4 chambers

History of the embryo heart, inside a pregnant woman:

Day of Development	The human heart beats about 100,000 times each day.
21	1 chamber
22	2 chambers
28	3 chambers
35	4 chambers

History of the Brain

God … you understand my thoughts from afar.

Psalm 139:2

Millions of years ago	Organism	Description
700	Anemone	Nerve net (no brain)
600	Flatworms	Brain appears
425	Fish	Tiny cerebrum (thought center) in brain
310	Reptiles	Small cerebrum 25% of brain area open to learning
35	Monkeys	Large cerebrum 50% of brain area open to learning
2	Humans	Largest cerebrum 75% of brain open to learning

History of the Face

Millions of years ago	Organism	Description
310	Reptiles	Jaw muscles open and close mouth.

100	Mammals	Face muscles appear. Mask - like.
35	Monkey	Eye and cheek muscles appear. Express greetings, anger and pain.
2	People	Lip and chin muscles appear. Can express a wide range of emotions with the face.

Development of a Human Face (after conception)

You [God] wove me in the womb of my mother.

Psalm 139:13

3 weeks	No face
1 month	Openings for the mouth, eyes, and nose appear.
7 weeks	Lips, jaw, and eyes appear.
3 months	Nose, ears, and chin appear. Face has a human look.

History of the Eye

Their eyes were opened, and they recognized him [Jesus]; and he vanished out of their sight.

Luke 24:31

Millions of years ago	Organism	Description
600	Flatworms	Eye spot. Light-sensing cells.
550	Sea snails	Pit eye. Light sensing cells in open hole.
425	Fish	Eyes enclosed, but do not move. 2D vision: Cannot see depth.
310	Crocodiles	Eyes move in socket.
35	Monkeys	3D color vision. Can see around objects.
2	People	3D color vision. Large brain to process images.

History of Legs

I [Jesus] say to you: 'Rise … and walk to your house.'

Mark 2:11

Millions of years ago	Organism	Description
385	Lung Fish	Thick fins at sides. Crawls onto land.
310	Lizards	True legs at sides. Zig-zag motion

| 230 | Mammal-like Reptiles | Legs are under body. Straight forward motion. |
| 2 | People | Back legs are under body for motion. Front legs are for "hand"-ling objects. |

Development of Human Legs (after conception)

God called me from the womb.

Isaiah 49:1

Day 33	Leg outgrowths appear. (No feet yet)
Day 35	Foot plates appear. (No toes yet)
Day 42	Toes appear, but webbed.
Day 43	Toes become separated.

History of Hands

He [Jesus] said … 'Stretch forth your [withered] hand.' And he stretched it out; and it was restored whole, like the other.

Matthew 12:13

Millions of years ago	Organism	Description
60	Tree Shrew	Claws on 5 digits

50	Tarsier	5 fingers with nails and pads
19	Old Word Monkey	Thumb appears, opposing fingers.
16	Ape: Chimp	Short thumb. Using simple tools.
2	People	Longer thumb. Using complex tools.

Development of Hands (after conception)

My bones were not hidden from you [God], when I was made in secret ...

Psalm 139:15

Day 28	Arm buds appear. (No hands yet)
Day 33	Hand plates appear — paddle shaped.
Day 41	Fingers appear together.
Day 52	Fingers separate. Nails appear.

History of Lungs

God says: 'I will make breath enter you, and you shall live.

Ezekiel 37:5

Millions of years ago	Organism	Description
385	Lung Fish	Swim bladder used as lung. Smooth breathing surface.
310	Reptiles	Larger lungs. Breathing surface is folded.
60	Mammals	Much larger lungs. O2 → In Complex breathing surface. Out ← CO2
2	People	Lungs with largest breathing surface (500 million air sacs in each lung)

One final thought. These "backward in time" connections contain many secrets. One example of this is the 2003 discovery by paleontologist Jack Horner of a fossilized Tyrannosaurus Rex in Montana. Horner's T Rex lived 68 million years ago. Within the fossilized thigh bone of this animal, Horner discovered a still elastic blood vessel! Scientists studied this rare find. They were able to decode 7 amino acid sequences from the collagen molecules in this T Rex vessel. To their great surprise, they found that 3 of the 7 amino acid sequences are also found in the domestic "chicken" of today!

ANTHROPOGENESIS

18. History of the Primates ...
Human Ancestors

Anthropogenesis is the study of human origins. Humans are classified as "mammals" because they are warm blooded, possess hair, give birth to their young alive, and provide their babies with milk. Humans also belong to a very special group of mammals known as "primates." Genetic studies have shown that primates began to appear on this planet about eighty million years ago. However, no primate fossils have been found before fifty-five million years ago.

Primates have several characteristics that are unique to them among mammals. Primate brains, in relation to their total body weight, are larger. While other mammals have digits with either claws or hooves, primates have flat nails on fingers and toes with fleshy pads. All primates, except humans, have curved big toes on their feet. This is to allow them to grasp branches while in trees. Most primates also have very dexterous hands. All old-world monkeys and apes (along with people), have thumbs that oppose their fingers, especially the "index finger" in humans. This, too, is an adaptation for life in the trees. The eyes of all primates face forward so that their visual fields overlap, making 3D vision possible. Many Primates also have color vision allowing them to recognize ripe fruit in the forest canopy. More advanced primates like humans have smaller teeth, including rounded molars, with enamel surfaces.

The earliest primates probably resembled the ring-tailed lemur of modern times. "Similodectes" was a lemur-like animal dating to about fifty million years ago. Lemurs have a short snout-like face. Their brain-body ratio is smaller than that of monkeys or apes. They move by hopping with their hind legs and communicate more with scents and vocalizations than with visual signals. Similodectes also had grasping thumbs and grasping big toes for arboreal life.

The earliest primate fossil presently known belongs to a tarsier found in 2002 in China. It dates to about fifty-five million years ago. Tarsiers are small creatures with very large hands, feet, and eyes. Their hands have long fingers with nails. They can leap over ten feet with their legs and see insects at night with their large eyes. In complexity of body structure, Tarsiers are somewhere between lemurs and monkeys.

The third step in primate development is the monkey. Monkeys have shorter snouts than lemurs. Monkeys are also fast learners, curious and crafty. Most species of monkeys are active during daytime, like humans. Monkeys possess tails, whereas apes do not (gibbons, orangutans, gorillas, and chimps are tailless). The New-World monkey has a grasping tail, whereas the Old-World monkey of Africa and Asia does not. Apes originated from Old World monkeys.

One of the earliest monkeys was "Aegyptopithecus" (also known as "Propliopithecus"). It was the size of a small monkey of today. Its fossilized remains were found in Fayum, Egypt, in 1965. It had a grasping big toe, forward-looking eyes, and a protruding lower jaw. Its dental structure resembles that of humans today. Its brain was smaller than any living monkey (the frontal lobes were especially small) and its teeth had little enamel, meaning that it probably ate the soft fruit in trees.

An old-world monkey known as "Proconsul" lived in East Africa's rift valley (which is in the present-day nations of Kenya and Uganda) about twenty-five million years ago. Proconsul was an ape-like monkey, a transitional form. One ape-like feature of this monkey was that it had no tail. However, it had the long back and curved hands of a monkey. Proconsul was discovered in Kenya in 1909.

The world's first fossil of a true ape was found in 2012 in the Rukwa Rift area of Tanzania by Nancy Stevens of Ohio University. The ape "Rukwapithecus" lived about twenty-five million years ago. It is estimated to have weighed only about twenty-six pounds.

The world's first fossil gorilla was discovered in 2007 in Ethiopia. It dates to about eight million years ago. The size of its teeth resembles that of gorillas of our time. This fossil was found by Gen Suwa of the University of Tokyo, who named it "Chororapithecus abyssinicus." This find importantly shows that Gorillas diverged from the human line about ten million years ago. Genetic studies further indicate that Chimpanzees took a separate path away from humans about eight million years ago.

19. Upright Walking Apes …
Freeing the Hands

Gorillas and chimps are "knuckle walkers." They do not walk upright due to their big toes being bent inward, together with other anatomical features. However, about six to seven million years ago certain apes began to walk solely on their back two legs. This had the advantage of freeing their front two limbs for "hand"-ling objects such as food. It also improved long distance running and eventually improved their hunting skills. Standing on two legs also increased their field of vision. Furthermore, studies have shown that up to four times less energy is used when moving forward on two legs rather than on four.

Upright walking gradually led to profound changes in anatomy. Legs over time became longer and arms became shorter. Knee and ankle joints also became stronger so as to support the increased weight. When walking on two legs, all of one's weight is alternately on one leg and then on the other. The curved big toe of earlier apes also moved into alignment with the other toes. And the pelvic bone became wider and bowl-shaped to stabilize the additional vertical weight. The hole where the spine enters the skull also moved from the side to the center of the head. Thus, the weight of the head was balanced on the spine and no longer required large muscles to hold the head up. This in turn allowed the skull to be thinner and the brain to expand in size.

One new problem that arose with these many changes was that the birth canal (the opening for a woman to deliver a baby) became smaller in size. This in turn limited the size of the baby's brain at birth. The result was that the child was less well developed at birth than many other creatures. (A horse walks in one day, a human child in one year) The brain therefore had to increase greatly in size outside the mother *after* birth. This made the baby heavily dependent on its mother for a much longer period of time. This in turn led to "pair bonding," or the long-term involvement of both parents in the development of their offspring.

The first bipedal, or upright walking ape, was "Sahelanthropus," whose fossil was discovered in 2001-2002 in Chad in central Africa. It is thought that this creature walked upright because the underside of its skull has a hole for the spine toward its center. However, its heavy brow ridges and face structure in general are very ape-like. Also, its brain case was only 350 cubic centimeters in volume, making its brain about the size of a chimp. Sahelanthropus walked on this planet about seven million years ago.

A second upright-walking ape who lived about six million years ago was "Orrorin," discovered in 2007 in Kenya. About twenty fossils of Orrorin have been unearthed. Its brain was also about the size of a chimp, but it had thick enameled teeth like humans, and legs where the bone buildup was typical of upright walkers.

"Australopithecines" were a transitional form between upright apes and humans. Australopithecines were man-like apes, fully clothed in hair. Their fossils were first discovered in 1924. One of the examples of these upright walking apes was "Ardipithecus," who lived about 5.6 million years ago. "Ardi" was a female discovered in 1994, with a partial skeleton also unearthed in 2009. Ardi was about 3 feet 11 inches tall and weighed about 110 pounds. Its fossil was discovered by Tim White, who is now at the University of California in Berkeley.

Australopith fossils from as many as 300 individuals have been discovered. An example of this large group is "Australopithecus afarensis" who lived between 3.8 and 2.9 million years ago. The most famous of this group is "Lucy," discovered in 1974 by Donald Johanson in Hadar, Ethiopia. Lucy was a three-and-a-half-foot-tall child, possessing a mix of ape-like and human features. She had an ape-like head with a jutting jaw, while her brain was about the size of a chimp. She had long arms and curved fingers for tree climbing. However, her pelvic, spine, leg, and foot bones revealed her to be an upright walker when on the ground.

Another group of Australopiths is "A. africanus." This is a group that lived between 3.3 and 2.1 million years ago. The fossil of one in this group is "Mrs. Ples," a middle-aged female upright walking ape, whose fossil was discovered within a cave in South Africa in 1947. Mrs. Ples had a small brain of 485 cubic centimeters, as compared to the average modern human brain of 1350 CCs (some would say 1400 CCs). Mrs. Ples was, however, an upright walker.

A 2015 study of hand bones in A. africanus revealed human-like metacarpals (the five bones between the wrist and the fingers) that allowed a strong grip between the creature's thumb and fingers. This kind of hand anatomy has only been observed in individuals where tools have been made and used. Thus M. Skinner (in *Science* magazine) argues that Australopiths may have been the world's first stone tool makers and users. Supporting this idea, in 1996 the fossils of another Australopith, "A. garhi," were found in Ethiopia, along with stone tools dating back 2.6 million years. Furthermore, a second Ethiopian site, at Bouri, has yielded some 3000 stone tools that date to 2.5 million years ago.

The next sign of the movement toward humans appeared in the creatures known as "Homo habilis" ("handy man"). Fossils of these creatures were discovered by Louis Leakey in 1960 at Olduvai Gorge in Tanzania. Later, in 1986, Donald Johanson and Tim White of the USA discovered the upper and lower limbs of Homo habilis, who lived

in Africa between 2.4 and 1.4 million years ago. Homo habilis had a brain somewhat larger than a chimp at 600-750
cubic centimeters. They also had a smaller face and smaller
teeth than were present in the earlier Australopiths. The
height of this creature was between three-and-a-half and
four-and-a-half feet. It also had a human-like foot and hand
bones that suggested that it could manipulate objects with
precision. Leakey also found stone tools with the fossils of
Homo habilis.

20. Humans Arrive ...
Hello!

The first true human fossils are thought to have been
those of "Homo erectus" ("upright man"), who lived between 2 million and 100,000 years ago. Homo erectus had a
brain size between 800 and 1100 cubic centimeters. Homo
erectus originated in Africa, but its kind eventually migrated toward Europe, India, and China.

One of the earliest examples of Homo erectus was "Turkana Boy," discovered in 1984 at Lake Turkana in Kenya.
This eleven-year-old boy lived about 1.6 million years ago.
Turkana Boy was discovered by Richard Leakey, the son of
Louis Leakey. Turkana's fossils include 108 bones, making
it an almost complete human skeleton. Turkana Boy still
had a low sloping forehead and strong brow ridges, but his
nose projected outward like modern humans, rather than

the flat nose seen in apes. In addition, his brain was 880 CCs, twice the size of the Australopiths. Furthermore, his legs were longer and his arms shorter, indicating that he walked erect. Stone tools were also found with his fossils, including the world's first hand axes.

One non-modern feature of Homo erectus fossils involves the spine. Its thoracic vertebrae were narrower than those in modern humans, probably meaning that it had less motor control over its thoracic muscles, which modify a person's breath to enable complex vocalizations. This ability is thought to be important in language formation.

A second example of Homo erectus is "Java Man," discovered in 1891 by the Dutch surgeon, Eugene Dubois. Java Man lived on the island of Java about 700,000 years ago. Java Man was about five feet eight inches tall. His thigh bone shows that he walked erect. He had small, human-like teeth, with a 900 cubic centimeter brain. Shell tools were found with him, probably used to cut meat.

A third example of Homo erectus is "Peking Man," discovered between 1921-1927 near Beijing, China. Peking Man lived in China about 750,000 years ago. By 1936, more than forty specimens, including six skull cases, had been found at the Beijing site. A large number of stone tools were also found. Peking Man's brain was comparatively large, between 1000-1100 cubic centimeters in size.

The next historical step in the climb to modern humans was "Homo heidelbergensis," discovered in 1907 near Heidelberg, Germany. This early man had a brain case nearly as large as modern humans: 1250 cubic centimeters, as opposed to 1350 in Homo sapiens. Heidelberg Man lived between 700,000 and 200,000 years ago.

Very few Heidelberg fossils were found at the 1907 site in Germany. However, in 1997 at Sima de los Huesos cave at Atapuerca in northern Spain, some 28 Homo Heidelberg individuals were found in a pit. These fossils date to about 400,000 years ago. There were roughly 7,000 fossil bones found at this site of Heidelberg Man. Geneticists were also able to extract and sequence some DNA from these bones.

Heidelberg Man had a flatter face than did Homo erectus. He also had a rounded dental arch very similar to modern humans (as opposed to the V-shaped arch of earlier man-like apes). Heidelberg Man used fire and wooden spears (eight have been found) to hunt and eat large animals. Some 500 stone spear tips were also found in Spain. Heidelberg Man is known to have built shelters or simple dwellings of wood and rock.

Some 27 complete limb bones of Heidelberg Man have been unearthed, making it possible to estimate his size. Male individuals were 5 feet 9 inches tall and weighed about 136 pounds. Females were 5 feet 2 inches tall and weighed about 112 pounds.

The next human population to consider are the "Neanderthal" people, who lived from about 400,000 to 40,000 years ago. Neanderthals existed alongside modern humans for many thousands of years. Neanderthal humans had a larger brain than modern humans. On average, the Neanderthal brain was 1500 cubic centimeters in size, as compared to 1350-1400 for Homo sapiens (modern humans).

Neanderthals were short in stature, with a robust skeleton and muscular build. Their forehead was flat and receding, and they had large eyes and a large nose. They were roughly five- to five-and-a-half-feet tall. The first Neanderthals were found in 1829, but not properly identified until fossils in the Neander Valley in Germany were unearthed in 1856.

Neanderthals originated in Africa, but about 200,000 years ago migrated into Europe, the Middle East, and parts of Asia. Neanderthals built hearths and were able to control fire for warmth, cooking, and protection from wild animals. Neanderthal sites also show that they regularly ate vegetables and wore animal hides for clothing. No "sewed" garments have so far been found. They are thought to have wrapped animal skins around their bodies and tied them. Neanderthals have been found living in caves, but also built open air shelters of wood branches and animal skins. In 2012 at a Neanderthal site in a cave in Gibraltar, an art-like engraving was found that resembled the grid lines for the game of Tic-Tac-Toe. There was also a bone "flute" found at Divje Babe in Slovenia. Neanderthals used blade tools similar to those used by Homo sapiens.

Recently geneticists have found the FoxP2 gene, which is related to language formation, in Neanderthal fossils. Furthermore, type O blood has been found in two Neanderthal males from Spain.

In 1957, eight Neanderthal adults and two infants were found in Shanidar Cave in northern Iraq. This site appears to have been a purposeful (religious?) burial, about 60,000 years ago. Fossilized pollen was found with the bodies, suggesting that flowers may have been present at the time. The adult skeletons also showed evidence of injuries that had healed, suggesting that Neanderthals cared for their sick and wounded.

The final step in this journey to modern humans is referred to as "Homo sapiens" ("Wise Men"). The earliest fossil evidence of Homo sapiens (our kind) are those found at Jebel Irhoud in Morocco beginning in 1961. In 2004, a team led by Jean J. Hublin of France discovered additional human skulls and many more human bones at the Irhoud site. These new human finds have been dated to 315,000 years ago. Previously, the "Omo fossils" found in Ethiopia by Richard Leakey in 1967 were the earliest known Homo sapiens. The Omo fossils were dated to 195,000 years ago.

While Homo sapiens originated in Africa, they have radiated outward covering the planet. Modern humans have a lighter skeletal structure than earlier human types. There is also a thin-walled skull with a flat, nearly vertical, forehead. The skull also has much less (if any) heavy brow ridges. The jaw is smaller and houses teeth which are considerably smaller than earlier types. Some researchers say that the average brain size of modern humans is 1400 cubic centimeters. Modern humans are also known to depend upon the use of language and complex tools to support themselves.

Another identifying characteristic of modern humans is "consciousness." The analytical level of consciousness seems to be considerably greater among modern humans than among other creatures. Consciousness at some level, however, seems to be present in a number of other non-human creatures.

In recent years the "mirror test" has been used to investigate whether a creature is conscious or not. The individual creature is put to sleep and then marked on a part of its body that it cannot normally see (for example, "the back"). When awakened it is then allowed to see its new mark in a mirror. If it then tries to remove the mark, it is assumed that "he or she knows" or is "conscious" of its being on its body. The various animals that seem to recognize themselves in the mirror are surprising. So far, magpies (who have no neocortex in their brain), the African Gray Parrot, elephants, orangutans, gorillas, chimps, and bonobos seem to have some level of consciousness. However, the depth or level of consciousness is an additional question. For example, it has been shown that some animals with consciousness can understand simple arithmetic, such as adding and subtracting. However, none of the above creatures have ever been seen doing algebra, trigonometry, or calculus! And none of the earlier primates investigated in this study have been seen doing architectural design! Cats and dogs are smart to be sure, but don't expect them to be studying at a technical school anytime soon!

21. Adam and Eve …
first human couple

In 1953, Francis Crick of England and James Watson of the USA authored a paper announcing the double helix structure of the DNA molecule. This "information molecule" is found in every cell in the human body. This recipe booklet tells how one specific human being is to be made from top to bottom. It is present from the moment of conception for each individual person. It also tells a scientist how the DNA molecular structure has changed over time. It lets the trained eye see the inside story of our own history.

In 1987, geneticists in the USA studied mitochondrial DNA in women. It had become known that mitochondrial DNA can only be passed on from mothers to daughters. From these studies, scientists concluded that all mitochondrial DNA in women alive today can be traced back in time to one woman who lived between 140,000 and 99,000 years ago. They gave this unknown woman the name, "Mitochondrial Eve."

Then similar studies were done in 1998 using the Y-chromosome in men. This chromosome is involved in maleness and is only passed on from father to son. Once again, the scientists were able to trace the various genetic components backwards in time to a single man who lived between 120,000 and 56,000 years ago. They named this hypothetical man, "Y-Chromosome Adam." Thus, it is possible that Adam and Eve lived sometime between 120,000 and 99,000 years ago!

The data concerning Adam and Eve is revealed in the Bible. There it states that God had placed Adam and Eve in what is called the Garden of Eden. In this beautiful part of nature, there is said to have been many things to see and to taste and to experience. God gave Adam and Eve free reign in this special place. Its trees and fruit represented all that makes life interesting and wonderful here on earth. Thus, it is seen that God intended humans be fully alive and happy on this planet.

God did, however, insist upon one area of understanding that humans were not to move into with their freedom. That was the sensitive area of the "knowledge of good and evil." Much later God would give humankind the Ten Commandments to help in this matter. Three of the Ten Commandments would outline a creaturely reverence for God. The last seven would provide rules to prevent a person from

hurting himself or herself and the people nearby. Taken together, the Ten Commandments are an abbreviated, yet
concise, definition of the nature of "love."

The world's first human sin occurred in the Garden of
Eden. The "Original Sin" was an act of arrogating to oneself
the ability to determine what is good or evil without refer-
encing God. In other words, the person alone determines
what is good or evil, without involving God. Put another
way, we become God ourselves, determining in our earth-
bound way the texture of reality, whether good or bad. Our
will becomes the criteria of good and evil, rather than allowing the will of God to occupy that space. "Don't tell me what
I can, or cannot, do!" That was the sin of Adam and Eve.
And it still is our sin!

Rebelliousness was not a new sin at the time of Adam
and Eve. In Heaven, a number of angels, led by Satan, had
turned against God (Revelation 12:1-10). Satan himself
wanted to be God (Matthew 4:9). If these evil-angels had
been allowed to remain in Heaven, Heaven wouldn't have
continued to be a "heavenly place." So, God used "removal"
as a form of treatment.

Originally, God had intended that the human species
would not die. His intention was that every human would
go directly from Earth into Heaven. If Adam and Eve had
followed God's directive, had not become involved with the
forbidden tree, they would not have died, but would have
gone straight into Heaven. In fact, the Bible itself speaks of

two such people who did go straight into Heaven! They were Enoch and the prophet, Elijah.

"Enoch walked with God, then he was not [anymore on earth] for God took him [to Heaven]."

Genesis 5:24

In the NT, this event is stated in the following words.

"By faith Enoch was changed so that he did not see death, and was not found, because God had taken him [to Heaven]."

Hebrews 11:5

Another man who was taken directly from Earth into Heaven was Elijah.

"It happened as they were walking and talking, that … they [Elisha and Elijah] were separated, and Elijah went up to Heaven in a tempest."

2 Kings 2:11

Eight hundred years later, in NT times, Elijah would appear with Moses before Jesus and three of Jesus's Apostles in what is called the "Transfiguration."

"Jesus took Peter, James, and John ... onto a high mountain ... and was transfigured before them. His face did shine as the sun and his garments became white as the light. And there was seen by them, Moses and Elijah talking with him [talking with Jesus] ..."

Matthew 17: 1-3

Thus, it can be seen in the example of Enoch and Elijah, what our life might have been like had we obediently followed God.

"God did not make death ... God formed man to be imperishable, the image of his own nature."

Wisdom 1:13 and 2:23

The question, however, might be asked: "Why has everyone on Earth been affected by the choice of Adam and Eve?" St. Paul answered this question in the following way.

"By one man [Adam] sin entered the world and death by sin [Genesis 3:4-6]; and so, death passed to all men, for all men [down to the present time] have sinned [in a similar way]."

Romans 5:12

In other words, death has come into the human world because of man's continuous rebellion against God.

"He forsook the God who made him, and lightly esteemed [God], the Rock of his salvation."

Deuteronomy 32:15

"They sin against you [God], there is no man who does not sin …"

1 Kings 8:46

"If we say that we have not sin, we deceive ourselves, and the truth is not in us."

1 John 1:8

The Adam and Eve event cannot be explained by merely saying that it was a simple error made by the innocent misuse of "free will." God, of course, created humans with free will, and by using that instrument, errors would inevitably be made. However, the Adam and Eve event had to do with willful rebellion. It was not the result of a simple error, and God treated it as such! This same kind of rebellious behavior is still a conspicuous part of the human story.

22. Food, Clothing & Shelter …
God's Help

Early peoples turned to nature to solve the various problems of getting food, clothing, and shelter. First, of course, was the need to acquire food to power their bodies. Earlier in time, such food had come from raw seeds, roots, wild fruits, and meat. Then about 1.7 to 1.5 million years ago the use of fire entered the scene. Charred animal bones from food preparation have been found in Swartkrans Cave near Pretoria, in South Africa, dating to 1.5 million years ago. Thereafter, heating and cooking became possible. Meat and vegetables became easier to eat and digest, therefore less time was spent on finding and preparing food each day. Softer food also provided more calories and nutrients for the hungry brain, possibly explaining why the size of the brain increased so rapidly in the development of early humans. Large grinding teeth also became less useful, so smaller teeth eventually replaced them. Lastly, cooked food brought people together. The fire and hearth experience, helped to create community. Stone hearths and clay fire pits, half a million years old, have been found in France, Hungary, and China.

Clothing was another challenge for early humans. Clothes provide protection from rain and cold. Modern humans, unlike their earlier ancestors, have very little body hair. Thus, the earliest form of clothing was made with animal hides having fur. A 2012 study reported that humans were wearing clothing by at least 170,000 years ago, and probably much earlier. Arctic peoples have, for many generations, made clothes from Caribou skins and footwear from waterproof seal skins.

There is no record of sewing among very early peoples. Dyed flax fibers have been found in a cave in the Republic of Georgia that date to 36,000 years ago. Sewing needles, made of bone, first appeared in use at Kostenki in Russia some 28,000 years ago. Textiles made by weaving, pictured on little pieces of hard clay, dating to 27,000 years ago have been found at Dolni Vestenial in the Czech Republic. And Venus figurines, that depict the use of sewed clothing, date to 25,000 years ago.

Shelter protects people from severe weather, shields them from insects and wild animals, and provides them with a place to sleep safely. A natural form of shelter used by early peoples were caves. Caves provided some protection from inclement weather, but less protection against wild animals. The first man-made shelters were made from animal skins stretched over a wood frame, with a small fire pit inside. The native American teepee was one such example, using pole-like branches leaning inward to a point and covered with buffalo skins.

The oldest known larger buildings are twelve 400,000-year-old huts at Terra Amata near Nice, in France. Found in 1960, they range from 26 to 49 feet in length and are between 13 and 20 feet wide. They were built by Neanderthal people using three-inch-wide wooden stakes braced by large stones and including a thatched roof. These huts also had built-in hearths or stone-lined fire pits.

In Bilzingsleben, Germany, Homo erectus also built three huts some 350,000 years ago. They were circular in shape, 9 by 13 feet across, using animal bones and stone in the construction. An elephant tusk was found in the center of one of these structures, thought to be its center post.

Eventually, raw earth came to be used in shelter construction. Wet clay-like earth was formed into bricks and then dried in the sun. When dry, the bricks were stacked to make an enclosure, or mud-brick house. Such mud-brick houses were built by native Americans in the southwest areas of the USA.

What's important to notice here is that early peoples, like modern humans, worked hard to make a life. Such survival efforts would provide the landscape for the later work of Christ, who would announce a special form of existence beyond this world and beyond death, namely, that of eternal life. Christ would demonstrate his divine power over death by way of his resurrection and a forty-day risen stay with the early Church community (Acts 1:3).

CHRISTOGENESIS

23. Christ Presence ...
"in his image"

The Lord continues to guide his creation. One-way Christ does this is by placing himself within the working environments of men and women across the globe. The creativity of modern humans evidences this ongoing presence of the Lord. In the areas of medicine, doctors are laboring to raise people from the dead. In the areas of transportation, people are becoming increasingly less bound in space. In the areas of communications, people are becoming increasingly less bound in time. And in the many kinds of factories that blanket the planet, people are creating a new world from such materials as metals, glass, and plastics.

All of these human efforts mirror the resurrection and ongoing life of Christ, and the fact that humans are created in his image (Genesis 1:26). Christ rose from the dead and was no longer space bound or time bound. He also preached continually about this new world above, namely, the Kingdom of Heaven. Humans on this planet, even those who give little attention to Christ, are inevitably doing his work. The whole planet has in a sense become an *echo* of Christ's presence.

Towards him we were running,
or from him we were running away,
but all the time he was in the center of things.

Karl Stern in Pillar of Fire

God is not space bound or time bound. God is not in space as we are, nor is he in time. God has no time! Humans are moving toward this new world; being closer to conditions of life in Heaven, where no one will be bound either by the limiting aspects of space or of time. And no one will ever again experience pain or death.

24. The God-Man …
"What sort of man is this?"

Mt 8:27

In the Trinity, God the Father's Word is Jesus, and the love between the Father and his Word is the Holy Spirit. God therefore is three subjects sharing one divine Nature.

The appearance of Jesus on this planet was very mysterious. When the angel Gabriel appeared to Mary, she was a virgin (Luke 1:34). The angel told Mary that the eternal God, rather than a man, would be the father of Jesus.

"The Holy Spirit will come upon you and …
the holy one born will be called Son of God (the Father)"

Luke 1:35

Jesus, the Father's Word, therefore, "became flesh and tented (temporarily) among us, and we observed his glory ..." (John 1:14) This did not primarily mean that God had become "a man," but rather that he had become "man," in the sense that he had seeded our world with his divinity.

We see this Heavenly orientation of Jesus for the first time at age twelve when he became separated from Mary and Joseph while in Jerusalem for the holy days. When they found Jesus in the Temple, he said to them: "Did you not know that into my Father's work it is necessary for me to be?" (Luke 2:49)

Another indication of Jesus's uniqueness as a divine Person is seen at the wedding feast at Cana. Here Jesus changed 120 gallons of water into wine to help the groom serve his guests during a weeklong wedding ceremony. During this event Jesus also addressed Mary in an unusual way: not as his "mother," but as "woman." "Woman, what is this concern [here] to you and to me?" (John 2:4) Later, while on the cross, Jesus would address Mary a second time, as "woman" (John 19:26). What Jesus was implying is that Mary is the new Eve, as Jesus is the new Adam (1 Corinthians 15:22). In Genesis, Eve is called "woman" eleven times before the sin of Adam and Eve. Only after their Fall does the name Eve appear.

Another indicator of Jesus as a divine Person is seen in the response given to some people who said to him: "Your mother and brothers are standing outside wanting to speak to you." To this, Jesus replied:

"Who is my mother, and who is my brother? …
Whoever does the will of my Father in Heaven
is my brother and sister and mother."

                                                    Matthew 12:50

In other words, the family of Jesus here on earth are those who show by their actions that they are on their way to Heaven. His family on earth are those who do the Father's will, and will eventually be with God the Father in Heaven

                                                    Matthew 6:10

A further indication that the God-Man's home was not primarily on this planet is the following statement. "Foxes have holes and the birds of the air have nests, but the Son of Man [me] has nowhere to lay his head" (Matthew 8:20).

The "powers" that emanated from Jesus were also a clear sign of his divine connection. One example of Jesus's un-earth-like power occurred on the Sea of Galilee, a body of water measuring 7 by 14 miles. A storm arose on this inland sea, causing fear to arise among his Apostles. Jesus, however, was not bothered by this. The Scripture says that he simply "admonished the winds and the sea, and there became a great calm" (Mt 8:26). This event was so direct that it shook the Apostles. They asked: "What sort of man is this that the winds and the sea obey him?" (Mt 8:27).

One of the primary concerns of Jesus, as the Father's Word, was to reveal God on this planet. Before Jesus arrived on Earth, gods and goddesses were worshiped all over its surface. That was because people did not really know who God was. They worshiped the sun, moon, planets, wild animals, and anything mysterious in nature. They did their best to respond to God's promptings. The Jews, however, were given a prophetic knowledge of God; but most people outside the Hebrew world knew little about it. So, God decided to reveal himself openly in the flesh.

In the beginning (of the universe) was the Word.
And the Word was with God (the Father),
and the Word was God (the Son) …
and the Word became flesh (in Jesus) and
tented (for a while) among us and we saw his glory

John 1: 1 & 14

God the Father was revealed on Earth by Jesus. Here are a few examples of Jesus's mysterious identity. "He who has seen me has seen the Father" (John 14:9). "I and the Father are One" (John 10:30). "The works (miracles) I do in my Father's name bear witness of me" (John10:25). "The words that I speak unto you, I speak not from myself, but the Father who dwells in me does his works" (John 14:10). "I am in the Father and the Father is in me" (John 14:11). "No one knows the Son but the Father, neither does any man know the Father except the Son, and he to whom the Son will reveal him" (Matthew 11:27).

Jesus, at one point in his ministry, asked his Apostles who they thought he was. After some discussion, Jesus said that Peter had answered correctly. Peter had said: "You are the Christ, the Son of the living God" (Matthew 16:16). Notice that in his definition Peter mentions "the Christ." The Christ had been revealed in the OT through Moses and the prophets. Jesus was "the Christ," the only person in the history of the world whose life story, his biography, was revealed before he came into the world. In fact, after his resurrection, Jesus explained this to two of his disciples:

And beginning from Moses [in the Law]
and in all the (OT) prophets,
he [Jesus] explained to them in all the [OT] writings
the things concerning himself.

                                                        Luke 24:27

25. Christ Raised People from The Dead

The NT records Jesus as having raised three dead people back to natural life during his public years. There may have been many more (Luke 7:22). He raised a synagogue leader's twelve-year-old daughter back to natural life (Mark

5:41-42). He raised the only son of a widow back to earthly life (Luke 7:12-16). And Jesus raised his friend Lazarus back to natural life, or life as we know it here (John 11:43-44).

Jesus also informed the twelve Apostles about his own impending death and resurrection. He told them three times before the events themselves occurred (Matthew 16:20 & 17:23 & 20:18). As a divine Person, Jesus knew of these events before their happening. Furthermore, Jesus never spoke of his own death without also speaking of his resurrection. "Destroy this temple (my body) and in three days I will raise it up" (John 2:19).

In my Father's house there are many dwelling places ...
I go to prepare a place for you ...
I will come again and take you to myself.

John 14:2-3

When Jesus had risen from the dead hundreds of people observed him. St. Paul speaks of this in his 53AD letter to the Corinthians.

"He was seen by 500 brothers at once; most of whom are still alive ... "

1 Corinthians 15:6

St. Peter says a similar thing:

"This Jesus God has raised up, of whom we are all witnesses."

Acts 2:32

"To us who ate and drank with him after he was raised from the dead."

Acts 10:41

This presence, with his many followers, was for an extended time.

"He presented himself alive to them by many convincing proofs, appearing to them during forty days and speaking about the Kingdom of God."

Acts 1:3

During his public ministry, Jesus cured untold numbers of diseases and medical conditions. These miracles of Jesus flowed from within him. He did not need to pray, like an OT prophet or a NT Apostle, when a miraculous action was needed.

[H]e went about in all Galilee teaching …
healing every disease and every illness among the people.

Matthew 4:23-24

[M]any followed him and he healed them all …

that might be fulfilled the things spoken through Isaiah …
"Behold, My Servant [Jesus] whom I chose."

Matthew 12:15-18
Isaiah 42:1

Whenever he [Jesus] entered villages or cities …
they put out their sick in the market places …
and as many as touched him were healed.

Mark 6:56

He [Jesus] said to them: 'Go report to John [the Baptist]
the things which you saw and heard: blind men see again,
lame men walk, lepers are being cleansed, and deaf men
hear; also, dead men are raised …'

Luke 7:22

This intervention of Jesus to cure and improve the condition of ordinary people was a sign of his divine presence and power. It was also a sign of the divine calling and work he expected of his followers. Humans were created in God's image (Genesis 1:26). Humans are to be an image of their creator. The medical community or medical profession is such an image. The idea of helping others who are sick or injured probably goes back to the earliest moments of humankind's existence on this planet. And it is still a considerable part of human life. In the following sections, we will try to document some of the more recent improvements in medical care.

MEDICINE
RAISING PEOPLE BACK TO LIFE

26. Discovery of Germs

The earliest evidence for a "magnifying glass" lens is found in a play by the Greek writer Aristophanes in the year 424 BC. In his play, *The Clouds*, one line mentions a magnifying lens being used to start a fire with kindling wood. Later, in the first century, Seneca (died 65 AD) wrote that with the magnifying glass he could read letters, "no matter how small or dim."

The magnifying glass is a lens that is thicker in its center and thinner on its edges. It can make things look 5 to 10 times larger than their real size. Earlier glass had many color imperfections, but by the first century AD, clear glass was available for lenses.

In the year 1306, a Catholic priest gave a sermon in Italy in which he said that it had been just twenty years since the eyeglass was invented. Therefore, it is believed by many, that about 1286 AD help for the eyes appeared in the form of a single lens for sight. However, the inventor is still unknown, and frames would not be developed until later.

About 1595 AD, Hans and Zacharias Janssen in the Netherlands put two lenses in a fixed tube, one lens at each end. The lens at the eye end, the "ocular lens," was usually ten power [10x]. At the other end, the lens nearer the object to be examined, was called the "objective lens." It could be, for example, twenty power (20x). Therefore, such a "Compound Microscope" would have two hundred power (10x20 = 200x). Objects would then appear 200 times larger than their true size!

By 1665, Anthony Leeuwenhoek, also of the Netherlands, had a 270 power (270x) microscope. With this instrument, he observed many new things in nature previously unseen and unknown. In 1673, he sent pictures of a bee's mouthparts to the Royal Society in London. In 1674, he observed green algae in water. In 1683, he observed bacteria from the plaque between his teeth. In 1698, he observed blood circulation in capillaries; and, in 1702, he observed single-celled organisms known as ciliates. He was the first person in all of history to see bacteria. However, for the next two hundred years the medical profession failed to connect bacteria with disease!

Also, in 1665, Robert Hooke of England saw plant "cells" for the first time with a microscope. They reminded him of the "cells" that Catholic priests live in while in a monastery. However, none of these discoveries concerning cells, bacteria, or single-celled organisms had any immediate effect on the theory of disease. It was not until about

1850 that doctors believed that many human diseases are caused by microorganisms or germs.

27. Germ Theory of Disease

It is now known that a human person has more microorganisms in their body than the number of his or her own cells! And most of these tiny organisms are the friends of mankind! For example, they help each person with their digestion, promoting a healthy GI tract. Some, however, cause various diseases. This section will look at a few key discoveries in this area of medicine.

In 1658, Athanasius Kircher, SJ, a Catholic priest in Rome, used a microscope to study fluids from patients who had died of Bubonic Plague. He saw the microorganisms that had caused the disease and reported it to the broader community. He also suggested effective measures to prevent the spread of this disease. He was ahead of his time. The medical community did not accept his findings.

In 1700, Nicolas Andry, a French physician, said that microorganisms, which he called "worms," caused smallpox and other diseases. Andry reported this in his book: "Orthopedie."

In 1810, Agostino Bassi, an Italian entomologist, began studying muscardine, a disease in silkworms. After twenty-five years of work, he announced in 1835 that the disease was caused by a microscopic parasitic fungus that would eventually be named in his honor: "Beauveria bassiana."

In 1856, M. Bigot asked the French scientist Louis Pasteur why the beer in his factory was souring. Pasteur examined the beer under a microscope. He not only found the spherical yeast cells used in beer making, but a rod-shaped microorganism, Acetobacter aceti. This bacterium was converting the alcohol in beer into acetic acid. This acid was causing the souring. In 1862, Pasteur found that a similar problem was threatening the wine industry in France. To solve this problem, he invented what came to be known as "Pasteurization." In this process, the wine was heated to 120-140 degrees F. All microorganisms in the product were killed in the heat. After several years, this process would also be used to eliminate pathogens in cow's milk.

In 1876, Robert Koch in Germany extracted bacteria from sheep who had died of anthrax. Koch then grew the bacteria and injected them into a mouse. The mouse developed anthrax and died. Using this method Koch isolated the bacteria that cause the human diseases of tuberculosis (TB) in 1882 and cholera in 1883. Robert Koch's methods led to the unraveling of many germ-caused diseases: typhus in 1880, tetanus in 1884, and plague in 1894.

In 1928, Alexander Fleming at St. Mary's Hospital in London discovered the world's first antibiotic, penicillin. Fleming noticed that the penicillin mold in his petri dishes was killing staphylococcus bacteria surrounding it. He reported this discovery but was unable to produce penicillin in great enough quantities for treatment. In 1940, Dr. Howard Florey of Oxford University and Dr. Ernst Chain from Germany undertook this project. In 1940, they injected fifty mice with the deadly staph bacteria. Half died, but those who had been given a penicillin injection lived. Florey and Chain (along with Fleming) shared the 1945 Nobel Prize in Medicine for their work on penicillin.

The problem then became how to produce enough of the penicillin. It took 2000 liters of mold culture to treat one patient with penicillin! Finally, with England at war, Dr. Florey decided in 1941 to fly to the USA to work on production methods. While in Peoria, Illinois, a lab assistant, Mary Hunt, arrived one day with a piece of cantaloupe covered with mold. By using cantaloupe as a culture, 200 times more penicillin could be produced than was possible with other methods. By 1943, penicillin's use in World War II had begun. As a result, only 1% of the soldiers died of bacterial disease in WWII. By comparison, in World War I, 18% of soldiers died of bacterial pneumonia.

Antibiotics are important because they make possible the killing of harmful bacteria within the body, without killing the person. There have been a number of different antibiotics produced since the discovery of penicillin. For example, in 1943 streptomycin was discovered by Albert

Schatz at Rutgers University. It has been especially important in treating TB. Other antibiotics in use include amoxicillin, cephalexin, erythromycin, and ciprofloxacin.

28. Treating Viruses

Viruses are too small to be seen with a light microscope. Viruses are parasites that have an RNA or DNA core surrounded by a protein coat. They are unable to reproduce without a host cell. They invade and take control of cells, and thereafter damage the body of any plant, animal, or human.

In 1796, Dr. Edward Jenner in England noticed that people exposed to cowpox did not suffer from smallpox, a dangerous human disease. He then decided to test his observation. He took some pus from a cowpox blister on a milkmaid's arm and scratched it into the skin of an eight-year-old boy. A single blister arose from the boy's scratch, but the spot soon healed. Then he inoculated the same boy with smallpox germs, and, predictably, the boy did not get the smallpox disease. Jenner's "vaccination" had worked, but Jenner himself did not know why.

In 1885, Louis Pasteur produced a vaccine for rabies by growing the rabies virus in the brain of a rabbit and then drying the affected nerve tissue. This drying weakened the rabies virus that had killed the rabbit. Pasteur then used the rabbit material to treat a boy, Joseph Meister, who had been bitten by a rabid dog and was near death. Pasteur injected the boy with the dried spinal cord matter of the dead rabbit. The boy lived. Yet Pasteur, like Jenner, did not know precisely why!

In 1884, Charles Chamberland in France developed a filter that had pores small enough to stop bacteria from getting through. Nonetheless, in 1892, Dmitry Ivanovsky noticed that something was getting through this bacteria filter. He thought it was a toxin. However, in 1898, Martinus Beijernck of the Netherlands thought differently. He believed the material getting through the filter was a new kind of infectious agent. He called it a "virus" (from late Middle English, meaning "slimy liquid, poison").

In 1903, the rabies virus was identified by Paul Remlinger in France, and, in 1909, Karl Landsteiner, an Austrian biologist, discovered the polio virus.

In 1923, Thomas M. Rivers of the USA revealed that viruses are parasites which need other cells to reproduce.

In 1927, scientists isolated the yellow fever virus in West Africa. Then Max Theiler, a South African microbiologist, developed the yellow fever vaccine in 1937 at the Rockefeller Institute in New York City. Since that time, over 400 million doses of this vaccine have been used against this disease.

In 1931, the electron microscope was invented by Ernst Ruska and Max Knoll in Germany. For the first time, viruses were able to be seen and studied in detail. In that same year, the influenza virus was also isolated from humans by Patrick Laidlaw in England.

In 1944, the first killed-virus vaccine for human influenza was developed by Thomas Francis in the USA. This effort was built upon the work of an Australian, Frank Burnet, who found that viruses lose virulence when cultured in fertilized hens-eggs.

In 1955, Jonas Salk, a student of Thomas Francis, developed the first polio vaccine. And, in 1981, the hepatitis B vaccine was approved for use. It was the work of the geneticist B. Blumberg and vaccinologist M. Hilleman of the USA.

29. Surgical Repairs …
extending life

Surgery cannot be done unless one understands the organ in question. Therefore, this section will look at the accumulation of knowledge involving specific organs and then examine what surgeons have done to repair them.

The Heart

The heart pumps 2000 gallons of blood each day.

Year	Scientist	Country	Scientific Achievement
1543	Vesalius	Belgium	Mapped the body's blood vessels.

1600	Fabricius	Italy	Studied valve flaps and proved that veins carry blood back to the heart.
1628	W. Harvey	England	Proved that the heart is a pump (not a hearth, or fire place). It pushes blood through arteries around the body.
1661	M. Malpighi	Italy	Found that tiny capillaries connect arteries to veins.
1707	L. Bellini	Italy	Found hard material inside the heart's own tiny blood vessels.
1842	C. Long	U.S.A.	Used ether to put patients to sleep. Long operations became possible.
1867	J. Lister	England	Developed germ-free, sterile, surgery.
1881	V. Czerny	Germany	Sutured the jugular vein in the neck.
1888	R. Mates	U.S.A.	Repaired a ballooning aorta blood vessel (aneurysm).
1900	K. Landsteiner	Austria	Discovered the human blood groups, making safe transfusions possible.
1906	J. Goyanes	Spain	Removed a ballooning artery and spliced in a vein from another location.

1912	J. Herrick	U.S.A.	Explained "heart attack" as a blood-clot that stops blood flow in the heart muscle.
1953	J. Gibson	U.S.A.	Developed the heart-lung machine. Puts oxygen in blood and pumps it around the body. Makes possible open-heart surgery.
1953	J. Gibson	U.S.A.	Closed a hole between the top two chambers in the heart.
1954	C. Lillehei	U.S.A.	Closed a hole between the bottom two chambers of the heart. CO_2 blood stopped mixing with O_2 blood.
1958	C. Lillehei	U.S.A.	Replaced the aortic valve in the heart with a man-made valve.
1963	Star & Edwards	U.S.A.	Replaced the mitral valve in the heart with a plastic valve.
1964	H. E. Garrett	U.S.A.	Performed the first successful "coronary artery bypass." Replaced a blocked coronary artery with a vein from the patient's leg.

The Lungs

The lungs put oxygen (O2) into the blood and remove carbon dioxide (CO2) from the body.

Year	Scientist	Country	Scientific Achievement
1600	R. Boyle	Ireland	Proved that air is needed for life. He put a candle and a bird in a sealed container. Then he pumped the air out with a vacuum pump. The candle went out and the bird died.
1731	S. Hales	England	Found the "Alveoli" air sacs in the lung. Each is about 1/100th of an inch wide. The lungs have about 300 million of them.
1771	C. W. Scheele	Sweden	Found two gases in air: 1/5th oxygen ("fire air"), and 4/5ths Nitrogen ("foul air").
1837	H.G. Magnus	Germany	Measured little oxygen in blue blood entering lungs, and much oxygen in the red blood leaving the lungs. Conclusion: Alveoli put oxygen into the blood.

1842	C. Long	U.S.A.	Used ether to put patients to sleep. This allowed long, complicated operations.
1883	R.U. Krönlein	Switzer-land	Was first to remove a cancerous tumor from the lung.
1895	W. Mace-wen	Scotland	Removed a whole cancerous lung from the body. Patient lived.

Muscles and Bone

The human body has 600 muscles. Muscles pull on bones.

Year	Scientist	Country	Scientific Achievement
Unknown			Wooden splint invented to stop the motion of an arm, leg, finger, or toe during healing.
1680	G. Borelli	Italy	Showed that muscles pull on bones.
1752	A. Von Haller	Switzer-land	Showed that nerves switch on muscles. When a nerve going to a muscle was cut that muscle would not contract.

1791	L. Galvani	Italy	Showed that muscles shorten when stimulated by electric current.
1842	C. Long	U.S.A.	Used ether to make painless surgery possible.
1883	J. Lister	England	Used iron wire to reconnect broken bone.
1886	C. Hansmann	Germany	Used an external plate to rebuild broken bone.
1893	W. Lane	England	Used long screws to bring broken bone pieces together.
1895	W.C. Roentgen	Germany	Invented the X-ray machine to see bones inside the body.
1924	A. Lambotte	Belgium	Used heavy nail-like wire inside arm and leg bones to reconnect pieces.
1940	A. Moore	U.S.A.	Did the first metallic hip replacement surgery.
1957	B. Walldius	Sweden	Did the first hinged artificial knee replacement. Used a cobalt, chromium, molybdenum alloy knee.

GI (Gastrointestinal) Tract

Year	Scientist	Country	Scientific Achievement
1842	C. Long	U.S.A.	Performed the first successful surgery using ether as anesthesia. Long operations without pain became possible.
1880	L. Rydygier	Poland	Removed one-fourth of a stomach with cancer. Reconnected small intestine to the remaining part of the stomach.
1888	K. Maydl	Germany	Removed whole colon with cancer. End of small intestine was connected to a hole in the skin. Stool came out of this hole.
1897	C. Schlatter	Switzerland	Removed all of a stomach. Small intestine was connected to the skin (stoma). Food was put into this exterior hole.
1935	Koenig	Germany	Invented a carrier plate and reusable bag to collect stool. Replaced pads and a belt.

Liver

The liver weighs three pounds. The liver's main job is to filter the blood coming from the digestive tract before it
passes to the rest of the body. The heart pumps one quart of
blood to the liver each minute.

Year	Scientist	Country	Scientific Achievement
1666	M. Malpighi	Italy	Proved that the liver makes one quart of bile each day. Bile breaks down fat and oil into fatty acids in the intestine.
1845	A. Kolliker	Switzerland	Found that red blood cells with a nucleus (made inside bones) are changed by the liver into oxygen carrying cells without a nucleus.
1848	C. Bernard	France	Discovered that the liver turns starch (glycogen) into sugar (glucose), the body's basic fuel. He found that the amount of glucose in the blood is regulated by the liver.

1888	C.J.A. Langenbuch	Germany	Removed the first liver tumor.
1888	H. Rex	Germany	Proved that the liver has two parts. Each of its two lobes has a separate blood supply.
1911	W. Wendel	Germany	Cut out the right lobe of the liver of a 44-year-old woman with cancer. The woman lived.
1932	H. A. Krebs	England/ Germany	Discovered the urea (or "Krebs") cycle: Toxic urea is excreted in the urine.
1963	T. Starzl	U.S.A.	Did the first liver transplantation. By 1985, 500 livers had been transplanted.

The Kidneys

Each kidney is about four inches long and has one million Nephrons. Nephrons remove one quart of urine from fifty gallons of blood each day.

Year	Scientist	Country	Scientific Achievement
1727	H. Boer-haave	Nether-lands	Discovered that urea contains nitrogen and is the solid part of urine.
1823	J.B. Dumas	France	Discovered that kidneys remove urea. He removed the kidneys from rabbits and cats. Result: there was a big buildup of urea in the blood of these animals.
1842	W. Bowman	England	Found the Nephron capsule, a two walled sac surrounded by tiny blood vessels that filters the blood.
1844	K. Ludwig	Germany	Proved that the "Bowman capsule" removes urea (with toxic ammonia: NH3) from the blood.
1862	J. Henle	Germany	Discovered the "Loop of Henle," a part of the Nephron that collects extra water and salts to be reused.
1869	G. Simon	Germany	Removed a diseased whole kidney from a patient. The person lived a normal life with the one remaining kidney.
1954	J. Murray	U.S.A.	Did the first successful human organ transplant. He

			put the kidney of one identical twin into the body of the other identical twin. Both lived a normal life.
1962	J. Hamburger	France	Began tissue matching for organ transplantation. Tissue of the donor should be a close match to that of the recipient.
1969	F.F. Borel	Switzerland	Borel, a microbiologist, discovered cyclosporine in soil, when he was vacationing in Norway.
1976	D. White and R. Calne	Switzerland	Studied cyclosporine for use in organ transplantation in rats. Today it is used for organ transplantation in humans. It prevents organ rejection.

The Pancreas

The pancreas produces a "juice" that assists in the digestion of protein, fat, and carbohydrates (wheat, rice, etc.) in the small intestine. The pancreas also produces insulin, which allows body cells to use the glucose present in blood. Glucose, a sugar, is the body's fuel.

Year	Scientist	Country	Scientific Achievement
1642	J. Wirsung	Italy	Found the tube connecting the pancreas to the small intestine.
1664	R. De Graaf	Netherlands	Collected pancreatic juice through a hole in the body.
1849-56	C. Bernard	France	Proved that pancreatic juice breaks down starch into sugar (glucose) and fat into fatty acids in the small intestine.
1867	Lister	England	Developed antiseptic (germ-free) surgery. Hand washing, sterile instruments and gowns became the new norm.
1869	P. Langerhans	Germany	Found the "islets" (clumps of special cells) in the pancreas that produce an unknown substance (insulin).
1874	W.F. Kühne	Germany	Found "trypsin" in pancreatic juice that breaks down protein into amino acids in the small intestine.
1882	F. Trendelenburg	Germany	Did the first removal of cancer from the pancreas.
1889	O. Minkowski	Lithuania	Discovered diabetes. They removed the pancreas from

	& J. von Mering		a dog and the dog got diabetes and died.
1921	F. Banting	Canada	Discovered the hormone "insulin" in the pancreatic juice of dogs. About 1in 50 cells in the pancreas produce insulin. Without insulin, glucose cannot be used as a fuel by the human body.
1966	W.D. Kelly and R. Lillehei	U.S.A.	Completed the first successful transplantation of the pancreas.

Skin Surgery

The skin is the largest organ in the body. The skin is a protective barrier against extremes of temperature and dangerous outside organisms.

Year	Scientist	Country	Scientific Achievement
1500 BC	Unknown	Egypt	Closed wounds by sewing open skin flaps together.
700 BC	Sushruta	India	Removed a skin tumor on the face. Then, sewed the skin flaps together.

| 1823 | C. Bunger | Germ-any | Developed skin grafting, moving skin from one part of the body to another. Using these methods, he reconstructed the nose of a 33-year-old woman. |

The Ears

A human ear consists of three parts: (1) The eardrum that collects the sound. (2) Three tiny middle ear bones that transmit the motion of the eardrum to the inner ear, or cochlea. (3) The cochlea is a 1 1/2 inch-long tube that contains 25,000 hearing "hair cells" moving in a fluid and connected to the brain by a nerve.

Year	Scientist	Country	Scientific Achievement
450 BC	Empedocles	Greece	Knew that the eardrum collects sound and vibrates.
1543	Vesalius	Belgium	Gave the first accurate description of the "mallus" and "incus" middle ear bones.
1546	G. Ingrassia	Italy	Discovered the middle ear "stapes" bone and the "oval window" into the cochlea.
1550	Fallopius	Italy	Discovered the "cochlea," the hearing organ, or inner

			ear. He also traced the auditory nerve into the brain.
1760	Cotumnius	Italy	Discovered the hearing hair cells within the cochlea fluid. Fibers near the oval window pick up high notes, further away from the opening, the low notes.
1850	Y. Yearsley	England	Repaired holes in eardrums by using a cotton-wool patch. Sound could then better move the eardrum.
1875	Kessel	Germany	Was first to unfreeze the stapes bone from the oval window. Hearing returned.
1942	E. Chain, H. Florey	England	Succeeded in producing the first antibiotic, penicillin. "Penicillin" kills "staph" germs that attack the middle ear.
1951	F. Zollner	Germany	Rebuilt an eardrum by using a thin slice of one of the patient's blood vessels.
1956	J.J. Shea	U.S.A.	Implanted the first artificial stapes bone and reconstructed the oval window by using a thin slice of the patient's blood vessel.

1957	A. Djourno	France	Implanted a man-made electronic cochlea for the very deaf.

The Brain

Year	Scientist	Country	Scientific Achievement
1595	Z. Janssen	Nether-lands	Invented the compound microscope.
1665	R. Hooke	England	Saw "cells' with his microscope.
1674	Leeuwen-hoek	Nether-lands	Examined the optic nerve of a cow.
1758	C. Hall and J. Dollond	England	Developed the Achromatic lens. This lens eliminated color fringes and improved the sharpness of microscopic images.
1837	J. Purkinje	Bohemia (now the Czech Republic)	Described the large nerve cells in the cerebral cortex.
1861	P. Broca	France	Proved that the left frontal area of the cerebrum, above the eye, is linked to speech production. This was the first localization of a brain function.

1870	G. Fritsch	Germany	Studied the "motor cortex" at the top of the brain which controls various movements of the body. He put dogs to sleep and then stimulated their exposed motor cortex. Precise points on the cortex caused precise movements of the dog's body.
1874	C. Wernicke	Germany	Found the area responsible for speech comprehension. "Wernicke's Area" is in the back of the left temporal lobe above the ear.
1888	W. Keen	U.S.A.	One of the first surgeons to successfully remove a brain tumor.
1921	O. Loewi	Germany	Proved that neurons can communicate across synapses by releasing chemicals. He discovered the first neurotransmitter, "acetylcholine."
1924	H. Berger	Germany	Developed the electroencephalograph to record the electrical activities of the brain.
1950	K. Lashley	U.S.A.	Found that memory has no single site in the brain.

				Memory is in the whole brain.
1974	Candace Pert	U.S.A.		Discovered the opiate receptor, or the bonding site for pain inhibitors in the brain.
1983	E. Kandel	Austria-U.S.A.		Discovered memory storage in neurons. Studied the sea slug. Found that short term memory consists of changes in existing synapses, while long term memory involves making new synapses. Found the CREB protein which increases the number of synaptic connections.

TRANSPORTATION
PEOPLE BECOMING LESS BOUND IN SPACE

30. People Becoming Less Bound in Space

Humans are becoming less bound in space and time —
like God. Humans are approaching life in Heaven. In
Heaven, people, like God, will be free to move about without various assists. We know this from the resurrection
movements of Jesus after he had risen from the dead (Luke

24:31-39); also, from the instantaneous movements of the Blessed Virgin Mary in Mexico (Guadalupe, 1531) and in France (Lourdes, 1858).

North America to Europe

Year	3,500 -mile trip from North America to Europe	Time to go from NY to London?
1492	Columbus sailed at 3 miles per hour	44 days
1929	"Queen Mary" ocean liner at 30 mph	5 days
1957	Boeing 707 Jet Airliner at 600 mph	8 hours
1976	"Concorde" Jet Airliner at 1300 mph	3 hours

New York to Los Angeles

Year	3,000-mile trip from New York City to Los Angeles	Time to go from NY to LA?
1800	Wagon pulled by horses at 5 miles per hour	50 days
1900	Trains going 50 mph	5 days
1936	Douglas DC3 Airliner at 200 mph	15 hours

1957	Boeing 707 Jet Airliner at 600 mph	5 hours

Conclusion: As carriers travel faster, it becomes easier to move among the continents. People are therefore becoming less space and time bound, thereby approaching conditions in Heaven.

31. Railroads …
extending people's legs and feet

Year	Inventor	Country	Invention
1550	Un-known	Germany	Mining cars on rails, pulled by mules, carrying metal ore.
1775	J. Watt	Scotland	Invented the modern steam engine with a rotary gear. The exterior condenser kept the steam cylinder hot at all times (saving energy).
1804	Trevith-ick	England	Developed the first high-pressure steam engine and built the first full-scale working railway steam locomotive.
1829	R. Ste-phenson	England	Built the "Rocket," the most advanced steam locomotive of its day. Traveled at 25 miles per hour.

1860	Civil War	U.S.A.	Top speed of trains: 60 mph.
1934	Budd Com-pany	U.S.A.	Pioneer Zephyr was a diesel-powered railroad train. Its top speed for travel between Denver, Colorado, and Chicago, Illinois, was 112 miles per hour.
1964	Hideo Shima	Japan	The "Bullet Train" uses a network of high-speed railway lines. Maximum speeds are between 150–200 mph.
1981	French National Railway Com-pany	France	Developed the TGV "high-speed train." It connects the main cities across France. Top speed: 357 miles per hour.
2004	China Railway	China	The "Harmony Express" is an electric high-speed train. Repulsive magnetic force allows the train to "fly" above its rails. Maximum speed: 236 miles per hour.

32. Automobiles …
personal extension of legs and feet

Year	Inventor	Country	Invention
1876	N. Otto	Germany	Invented the 4-stroke gasoline engine. (1) Intake stroke brings gas in (down piston); (2) Gas and air are then compressed (the up stroke of the piston); (3) At top a spark causes the gas to explode, pushing the piston down (the power stroke); (4) Finally, the exhaust in cylinder is removed as the piston moves up. This 4-stroke cycle is then repeated over and over again.
1885	K. Benz	Germany	Built the "Motor-wagon": a 1-cylinder, 4-stroke gas engine in a 3-wheel "open carriage." Top speed: 8 mph.
1886	G. Daimler	Germany	Built the "Daimler" motor car: a 1-cylinder 4-stroke gas engine in a 4-wheel "open carriage." Top speed: 10 mph.
1896	H. Ford	U.S.A.	Built the "Quadricycle" or cart on four

			bike tires powered by a 2-cylinder, 4-stroke, ethanol engine. It had two forward gears, no reverse. Top speed: 20 miles per hour.
1903	H. Ford	U.S.A.	Built the "Model A Ford": a 2-cylinder, 4-stroke, 10 horsepower gas-engine. Hand brakes. Top speed: 30 mph.
1908	H. Ford	U.S.A.	Developed the "Model T Ford," the first affordable car for the common man. This car had a 4-cylinder, 4-stroke, 20 horsepower, gas engine. Two forward, one reverse gears. Top speed: 45 mph.
1913	H. Ford	U.S.A.	Began the "assembly line" form of production, using interchangeable parts. By 1924 the 10 millionth Model T Ford had rolled off the assembly line in Detroit, Michigan.

33. Auto Improvements

Year	Improvement
1903	Open body. Hand-crank starter. Steering Wheel. Windshield. Foot accelerator. Speedometer. Hand brake. Electric lights. Horn.
1910	Closed body. Electric starter. Hydraulic brakes. Rear brake lights. Rear view mirror.
1920	Heater. Windshield wipers. Balloon tires. Built-in radio.
1940	Air-flow body. Turn signals. Automatic transmission. Tubeless tires. Seat belts.
1960	Power steering. Air conditioning. Plastic body parts.

34. Airplanes …
moving people upward from the Earth

The airplane has wings that curve on the top side while being straight on the bottom side. As air passes over the top, curved surface, it becomes less dense. However, the air going across the bottom surface of the wing remains high in pressure. This higher-pressure air pushes up on the wing, so the plane rises.

Year	Inventor	Country	Invention
1894	Lilienthal	Germany	Developed the "hang glider." He made 2000 research flights to perfect his design.
1896	O. Chanute	U.S.A.	Built a two-wing, bridge-like glider near Chicago. It had fixed tail fins, but no engine. The Wright brothers watched him fly.
1903	Wright Brothers	U.S.A.	Built the "Wright Flyer." It had two wings, a 12-horsepower aluminum gas engine, and an elevator and rudder. The pilot twisted the spruce-muslin wing to roll left or right. It made the world's first powered flight at 31 mph.
1908	S. Cody	England	Built the "Aeroplane No. 1" with the world's first "ailerons" to control roll.

1909	L. Bleriot	France	Built the first powered monoplane (with one rather than two wings). The Bleriot X1 was flown across the English Channel in a 36- minute flight. With a 25 horse power engine, it flew at 45 mph and landed in England on its own wheels.
1919	H. Junkers	Germany	Built the world's first all metal transport plane. The Junkers J13 had the first "cantilever wings," or wings that use no external struts or bracing. Carried four passengers. Was powered by a 6-cylinder, 158 horse powered engine. Maximum speed 107 mph. Ceiling:16,000 feet. Range; 870 miles.
1926	E. Stinson	U.S.A.	Built a six seat, high winged monoplane. The Stinson SM-1D

			had a 220 horse power engine, cabin heating, wheel brakes, and an electric engine starter. Maximum speed: 135 mph. In 1928 "Northwest Airways" began to use this plane to fly passengers.

35. Airliners ...
cloud travel

Year	Inventor	Country	Invention
1936	Donald Douglas & Arthur Raymond	U.S.A.	The Douglas DC 3 was a fixed-wing propeller-driven airliner. Its cruising speed (207 mph) and range (1,500 miles) changed air travel around the planet. An all-metal plane, it carried 32 passengers. Crossed the 3000 miles of the USA in 15 hours with 3 refueling stops. Had two 1250 HP piston

			engines. Over 11,000 were built.
1952	Comet	U.K.	World's first jet airliner. It carried 40 passengers at 460 mph. Range: 1,500 miles. Number built: 114.
1957	Boeing 707	U.S.A.	Jet airliner that carried 175 passengers at 627 mph. It had sweptback wings with 4 podded jet engines. Range: 5,754 miles. Number built: 1011.
1967	Boeing 737	U.S.A.	Medium-range, narrow-body, twin jet. Passengers: 150-200. Speed: 583 mph. Number built: 9,500. Still in production.
1969	Boeing 747	U.S.A.	A wide body, jumbo jet airliner. Four turbofan jet engines. Passengers: 400. Speed: 614 mph. Range: 8,350 miles. Front loading cargo door. Number built: 1,500.

1976	Concorde	U.K. & France	The Concorde was the first successful supersonic jet airliner. Passengers: 125. Speed: 1350 mph. Range: 4500 miles. Number built: 20
2009	Boeing 787	U.S.A.	The "Dreamliner" is a wide body, twin engine jet airliner. Passengers: 250-300. Range: 9,500 miles. Speed: 587 mph. Is 20% more fuel efficient than other like-sized airliners, due to its carbon-composite body. Number produced thus far: 550

36. Helicopters …
rising up, ascending – like Jesus

Acts 1:9

Year	Inventor	Country	Invention
1923	J. Cierva	Spain	Built the "Autogiro." Its 16-foot rotor blades were not power driven. A front-mounted gas engine with a propeller pulled the plane forward. Hinged blades prevented roll-over. With rotor blades turning, the plane could take off and land on a short runway.
1936	H. Focke	Germany	Built the world's first helicopter, the Focke-Wulf FW-61. Two 23-foot rotors were power-driven by an internal gas engine. There was no rotor on the tail. Takeoff was straight up, where it could hover,

			and then land straight down. Number built: only one.
1942	I. Sikorsky	U.S.A.	Built the world's first single rotor helicopter, the XR-4. It had two seats, a single 3-bladed rotor, powered by a radial gas engine. It had a tail rotor that was powered. Speed: 75 mph. Ceiling: 8,000 feet. Number built: 131.
1945	A.M. Young	U.S.A.	Designed the "Bell 47," a multi-purpose, two passenger helicopter. It had a 6-cylinder, 280 HP piston engine. Speed: 105 mph. Range: 245 miles. It had a plastic bubble front where the passengers sat. In 1949 it was the first helicopter to fly over the Alps Mountains. Number built: 5,600.

| 1959 | Sikorsky Aircraft Corp. | U.S.A. | A rescue helicopter that could land on water. Employed a rescue basket on a cable. Used by the US Coast Guard for air-sea rescue. Its 3-blade rotor was powered by one turbo-shaft engine. Speed: 109 mph. Range: 412 miles. Ceiling: 11,200 feet. |

37. Rocket Travel …
humankind moving outward, toward Heaven

Year	Inventor	Country	Invention
1926	R. Goddard	U.S.A.	Built the first liquid rocket. It used liquid oxygen with gasoline fuel. Height: 10 feet. Width: 1 foot. Speed: 64 mph. Range: 184 feet into the air.
1942	W. von Braun	Germany	Built the world's first long range rocket, the "V-2." Operational range: 200 miles. Liquid oxygen and ethanol as the fuel. Height: 46 feet.

			Width: 5 feet. Speed: 3,580 mph.
1967 -73	NASA	U.S.A.	"Saturn 5" moon rocket. Three stages. Stage 1: Liquid oxygen with kerosene fuel. Stage 2 and 3: Liquid oxygen with hydrogen fuel. Height: 363 feet. Width: 33 feet. Speed: 24,790 mph. Payload: 1 moon car and three men. Range: To the moon and back six times.

Conclusion: Humans are moving outward from this planet, toward Heaven.

COMMUNICATIONS
MAKING PEOPLE LESS BOUND IN TIME

38. People Becoming Less Bound in Time

Communications involve many areas of human endeavor: Sign language, speech, writing, books, libraries, telegraph, telephone, still photography, the phonograph, motion pictures, radio, television, sound and video recording, the computer, satellites, internet, and smart phones.

Three examples from the above list stand out.

First, the **phone**. One can push a few numbers on one's smart phone and in a few seconds be talking from Los Angeles to someone in New York City, about 3000 miles away. The conversation is almost instantaneous and the connection time but a few seconds. If a satellite were added to this mix, one could talk to another person in Tokyo, Japan – half way around the planet. Such an instantaneous connection would pass through many "time zones" and seem as if the person was immediately before us. Lastly, if a computer was used to make such a call, a program like "Skype" would allow not only a voice contact, but also live pictures of the person with whom one is talking.

Motion pictures also make time stand still. For example, one can watch Humphrey Bogart in the movie *The African Queen* even though Mr. Bogart was called away to the Lord many years ago. T*he African Queen* was made in 1951. By watching a motion picture, one experiences time as if it were "standing still." Since humans have been made in God's image (Genesis 1:26), it is not surprising that we can experience time in a similar way. In God, there is no time. Movies provide a similar experience.

Lastly, there is **television**. It, too, breaks the bonds of time. For example, one can see the Pope speak at the Vatican in Rome, and it will be instantaneously seen and heard around the planet. The Pope can be seen and heard in all time zones at once because the broadcast is "live"—it is transmitted through the use of communication satellites.

Another such example comes from the space program. In 1971, the whole planet watched astronauts driving a car on the moon while it was happening! This was an amazing event, broadcast "live" from a planetary satellite!

In these ways, communications can override both space and time, giving people here on Earth a glimpse into one's future life in Heaven.

39. Writing …
making time stand still

People can communicate "bodily" with other humans in various ways. People can shake hands, hug, or, when angry, pound on a table. People can also point to something, and raise or lower one's voice, to be understood. Hand signs can also be used to speak with one another, what is commonly referred to as "sign language."

When speech developed among humans, communication became more precise. Words were used in a wider range of ways, than is possible with sign language. There were also many more nouns and adjectives available to express oneself. However, speech could not stop time, or preserve the communication of one moment, for use at another time. That became the job of the next great advance in communications, namely, writing.

At least four early systems of writing were developed on this planet. (1) Cuneiform writing in Sumer between the Tigris and Euphrates Rivers in modern-day Iraq. 2) In Egypt with its hieroglyphics system (3) In China with its many picture symbols. And (4) in America with the Mayan and Incan writing systems.

Writing provides a tool for recording information uniformly and in greater detail than is possible with the spoken word. Writing allows societies to transmit information and also to share and accumulate knowledge.

Sumerian cuneiform writing on clay tablets and Egyptian hieroglyphic writing on papyrus are considered to be the earliest writing systems. Both of these systems emerged between the years 3,400-3,100 BC.

40. Books & Libraries …
making thoughts timeless

A book can put the reader inside the mind of a person who may have lived on this planet a thousand years ago. Across the centuries, such an author will speak clearly but silently inside of one's mind. The bonds of time are thereby broken in this mysterious way.

Early writers recorded their thoughts on clay tablets in Mesopotamia or on papyrus in Ancient Egypt. Papyrus was a paper-like material made from the papyrus plant that grows along the Nile River. Papyrus sheets could be connected together into long rolls called "scrolls." Some scrolls might be thirty feet long and were read from left to right. Ancient Egypt, Greece, and Rome all used scrolls to record business records, event history, and stories.

In the Near East, between the Tigris and Euphrates Rivers, the preferred writing material was clay. Wedge marks were made on wet clay to express thought. This kind of writing was called cuneiform. It was mostly used for government records, business, and inventories. Long stories, national epics, and other writing challenges were too difficult or too long to put effectively on clay.

Libraries were constructed to hold collections of scrolls. In the ancient world, libraries were often connected to a temple, the king's palace, or a school where scribes were being trained. Private book collections were rare. The first books (or scrolls) were expensive because all books before the invention of printing had to be hand copied.

In the first century, the century of Christ, the paged book or "codex" was invented. The codex was a "bound" book which consisted of individual pages attached to each other on one side and held together with boards or cloth.

The codex "book" eventually replaced the scroll in the Roman world. Today, libraries around the world are filled with "books" rather than scrolls.

Many ancient libraries have been destroyed by war or neglect. The following are a few ancient examples that have been unearthed.

2700-2400 B.C.	Nippur was among the earliest of the Sumerian cities. It is presently located in the nation of Iraq. The Temple Library of Nippur was excavated between 1889-1900 by a team from the University of Pennsylvania. Since 1889 some 40,000 Cuneiform texts have been recovered.
2500-2250 B.C.	Ebla was one of the earliest kingdoms in today's Syria. The site is Tell Mardikh, about 34 miles from Aleppo in Syria. It was excavated between 1964-2011. The Ebla library was found in two rooms near a large audience hall in the king's palace. Room one had government and economic texts. Room two had epic narratives, myths, and school related topics. In all, over 2000 cuneiform texts were recovered.
2000 B.C.	The Library at Thebes, in Egypt, was located about 500 miles south of the Mediterranean Sea along the Nile River. It was built about 2000 BC. Above its front door were the following words: "Medicine for the Soul." This library, however, was destroyed. Nonetheless, one can get a sense of its contents by examining the story "Sinuhe," what some believe to be the

	world's first "novel." "Sinuhe" is a narrative set after the death of Pharaoh Amenemhat 1 (ruled 1991-1962 BC). It is believed by scholars that this story was written shortly after the Pharaoh's death. Sinuhe is an official who travels with Prince Senwosret to Libya, where he hears of the Pharaoh's death. He decides not to return to Egypt, marries, has sons, defeats a powerful opponent, and becomes an old man. Then he receives an invitation from King Senwosret to return to Egypt where he completes his life in royal favor.
650 B.C.	The Library of Ashurbanipal (ruled 668-627 BC) was located in Nineveh. Today, the site of Nineveh is close to Mosul in northern Iraq on the Tigris River. It was then the capital of the Assyrian Empire. This library of over 30,000 texts can now be seen in the British Museum in London. In the Hebrew Bible, Ashurbanipal is called Asenappar (Ezra 4:10). This library contained hundreds of medical texts and the Epic of Gilgamesh. It was discovered in 1897 by the British archaeologist A. Layard.
117 A.D.	The Library of Celsus in Ephesus (a city in the Roman Empire) was begun in 114 and finished in 117 AD. One can still see the standing library at the archeological site in Selcuk, Turkey. The library stored more than 12,000 scrolls.

41. History of Electricity …
extending the energy of people

Year	Inventor	Country	Invention
1663	O. Guericke	Germany	Built the first reliable "spark"-making machine. Held a cloth against a spinning ball of sulfur.
1729	S. Gray	England	Was the first to experiment with electrical conduction. Found that wire will carry a spark 800 feet. Thus, electricity is not "static," but rather "flows" like a fluid.
1781	L. Galvani	Italy	Discovered that the legs of a dead frog twitched when touched by two different metals. Galvani thought this was due to the life

			force within the frog.
1799	A. Volta	Italy	Studied Galvani's experiment. He thought the electricity was in the metal circuit. He stacked thin copper and zinc plates together. Then he added, between each plate, a piece of cardboard soaked in salt water (to represent the frog's leg). He made six sets of these three items and connected them in a circuit. The result: the world's first "battery," producing a "current" of electricity.
1819	H. Oersted	Denmark	Found that a compass needle is pulled toward a

			wire that is carrying an electric current. In other words, a wire carrying an electric current creates a magnetic field around itself.
1831	M. Faraday	England	Found that breaking the lines of force in a magnetic field will create an electric current. He wound an open coil of wire and connected it to a galvanometer to measure any current. Then he moved a bar magnet back and forth through the wire coil, inducing an electric current in the wire. This was the world's first "generator" of electric current.

			Faraday then made a "disk generator": a circular copper disk rotating between the poles of a horseshoe magnet. It produced a small direct current (DC) voltage.
1867	C. Wheatstone & W. Siemen	England & Germany	Built the first practical generators for industry. Both used electromagnets (many turns of wire in a coil) rather than permanent magnets. This greatly increased the power output of generators.
1873	Z. Gramme	Belgium	Built the first electric DC motor, powerful enough for use in industry. Earlier motors had been

			mostly used as toys.
1879	T. Edison	U.S.A.	Invented the first practical electric light bulb. In 1882 Edison built a DC generator in one part of New York City. It supplied electricity for 14,000 lights in 900 buildings over a one square mile area. One problem: DC generators can send out electricity for only about two miles from the generation site.
1883	N. Tesla	Croatia	Invented the AC (alternating current) motor. Most motors used in the world today are AC motors.
1888	N. Tesla	Croatia, U.S.A.	G. Westinghouse bought Tesla's AC

			motor and generator patents. Unlike DC power, AC can be sent over long distances. An AC voltage can be stepped up by a "transformer" at the generator site and sent many miles, and then stepped down with another transformer at the user site. This cannot be done with DC (direct current) instruments.
1895	Tesla & Westinghouse	U.S.A.	Built the Niagara Falls AC power plant. Sent out electricity to Buffalo, New York, 17 miles away.
1911	W. Coolidge	U.S.A.	Invented the tungsten light bulb. It used a

			tungsten metal filament. All air was removed from the glass bulb and was replaced by a non-reactive gas. Its light was equal to 105 candles.
1954	Chapin, Fuller & Pearson	U.S.A.	Daryl Chapin, C. Fuller, and Gerald Pearson at Bell Labs built the first silicon solar cell to make electricity from sunlight.

42. The Telegraph …
instantaneous messages across a continent

Year	Inventor	Country	Invention
1799	A. Volta	Italy	Invented the first battery, making it possible to produce a "current" of electricity.
1825	W. Sturgeon	England	Invented the "electromagnet." When an electric current flows in a

			wire, it creates a magnetic field around that wire. By winding more wire into a coil, the magnetic field is strengthened. Its magnetic strength is controlled by the amount of current allowed to flow through the wire.
1832	S. Morse	U.S.A.	Imagined the telegraph. A wire connects two locations. Vary the electricity at one end and that signal will be repeated at the other end. Use dots and dashes to identify each letter of the alphabet. A dot would be a small amount of electricity. A dash would be a larger

			amount of electricity. Both ends would have a tapping device. For example: The letter "e" is one dot. The letter "a" is one dot and one dash. The letter "s" is three dots … and so forth.
1835	S. Morse	U.S.A.	Morse built the one-wire telegraph. However, he did not have enough money to string copper wire from one city to another. Not until 1843 would Congress provide money to build a telegraph line from Washington DC to Baltimore, Maryland.
1837	Cooke & Wheatstone	England	Patented a telegraph using 5 wires to deflect 5

			needles indicating the letters of the alphabet. They went on to build the first commercial line in 1838.
1844	S. Morse *	U.S.A.	The 44-mile telegraph line from DC to Baltimore was completed. Morse sent the first message: "What hath God wrought?" By 1854 there were 23,000 miles of telegraph lines in the USA.
1858	C. Field	U.S.A.	Laid the first undersea telegraph cable from North America to Europe. It operated for only three weeks. However, by 1866 a new cable was a success.
1861	Western Union	U.S.A.	Completed the first telegraph line

			across the whole USA.
1874	T. Edison	U.S.A.	Invented the quadruplex system. It allowed for four messages to be sent simultaneously on one telegraph line. By 1915 many messages could be sent on one line at the same time (called multiplexing).

43. The Telephone …
a person's voice flies, is not space bound.

Year	Inventor	Country	Invention
1849	A. Meucci	Italy	Began designing a "talking telegraph" or telephone in 1849. He used it to talk from his basement lab to his invalid wife on the second floor of their house.

| 1856 | A. Meucci | Italy | Meucci's notebook described the telephone: "It consists of a vibrating diaphragm (collecting sound) and an electrified magnet with a spiral wire that wraps around it (the electromagnet). The vibrating (sound) diaphragm alters the current of the magnet. These alterations of current are transmitted to the other end of the wire (the receiver), creating like vibrations in the receiving diaphragm that reproduces the words (in a speaker)." |
| 1870 | A. Meucci | Italy | Meucci's telephone transmitted a voice message one mile over copper wire wrapped in cotton. |

1871	A. Meucci	Italy USA	Meucci applied for a US patent but failed to include the word "electro-magnet "as its central feature. His application failed, and because of his sick wife he was too short of money to resubmit the application.
1876	A.G. Bell & E. Gray	U.S.A.	Alexander G. Bell, a speech professor at Boston University, and Elisha Gray, a science instructor at Oberlin College, both applied for a telephone patent on the same day in 1876. Apparently, Bell was first because he received the patent for his telephone.
1879	Lowell, Mass.	U.S.A.	"Phone numbers" were first used in

			Lowell, Massachusetts. Previously a switchboard operator was given the name of the person that the caller wanted to speak with. The operator then connected the two.
1891	A. Stowger	U.S.A.	Invented the "dial phone." However, it was not introduced until 1904.
1973	M. Cooper	U.S.A.	Martin Cooper of Motorola invented the world's first hand-held mobile phone.
2002	Congress	U.S.A.	The U.S. House of Representatives acknowledged that A. Meucci was the first inventor of the telephone.

44. Still Photography …
stops time, a sign of eternal life.

Year	Inventor	Country	Invention
1667	Unknown		Artists used a light tight box with a convex lens and an inner mirror angled upward, projecting an image onto a flat glass screen at the top. This image was then traced onto paper by artists.
1727	J. Schulze	Germany	Studied "silver nitrate" which darkened over time. He placed the silver compound on a plate and covered part of the plate. The uncovered part turned black in the sun, while the covered part did not, proving that the sun's light (and not

			the sun's heat) was the cause.
1777	C. Scheele	Sweden	Studied "silver chloride." He found that in light it turned black and that the "black" was the metal silver. Silver chloride + light → Silver + chlorine
1816	N. Niepce	France	Made the first photograph using a small camera and a piece of paper coated with "silver chloride," which darkened when exposed to light. However, he had no method of stopping the darkening process.
1837	L. Daguerre	France	Developed the first practical photo process. He treated a silver plate with

			iodine vapor to give it a coating of light sensitive "silver iodide." After exposure in a camera, he "developed" its image, using mercury vapor and a solution of table salt. There was only one drawback: each picture was separate. Copies could not be made.
1840	H. F. Talbot	England	Invented the first "negative" to "positive" photo process, called the "Calotype." He used "silver iodide." Camera time: 1-2 minutes. He then stopped the darkening process with gallic acid. The "negative" could then be used to make prints.

1871	R. Maddox	England	Invented the "dry film" process. Put "silver nitrate" into a "gelatin" form as an emulsion. This dry film process allowed film to be manufactured for the first time and stored.
1888	G. Eastman	U.S.A.	Introduced celluloid film (a kind of plastic) and the "Kodak" camera. It was preloaded with a roll of film for 100 photographs. When exposed they were sent to a factory for developing and the camera was reloaded there.
1913	Herbert & Huesgen	U.S.A.	Made the first 35 mm camera in New York. Called the "Tourist Multiple." Used a 35 mm roll of film. Camera

			was light weight and strong.
1935	L. Godowsky	U.S.A.	Created the first practical color film: Kodachrome. Its emulsion had three layers: one for red, one for green, and one for blue light.
1969	G. Smith & W. Boyle	U.S.A.	Smith and Boyle developed the CCD (charged couple device) at Bell Labs. This CCD image sensor is the heart of the "Digital Camera." Digital cameras do not use film, but capture photographs on an internal memory chip or digital memory card.
1970	Fairchild	U.S.A.	Fairchild Imaging built the first commercial CCD imager. It had a 10,000- pixel sensor.

1975	S. Sasson	U.S.A.	Invented the first self-contained digital camera at Eastman Kodak. It weighed 8 pounds and used the CCD image sensor chip. Was produced for military and scientific purposes only.
1981	Sony	Japan	The Sony Mavica was the world's first electronic camera for sale to the public.

45. Sound Recording …
spoken words are no longer time bound.

Year	Inventor	Country	Invention
1877	T. Edison	U.S.A.	Invented the tinfoil phonograph or "talking machine." Two parts: (1) a turning cylinder covered with tin foil (2) a cone-shaped horn to collect sound with a needle attached to it. One spoke into the horn and the needle

			etched a sound tract onto the turning tinfoil. To play back the sound, one placed the needle at the beginning of the tract and turned the cylinder.
1877	D. Hughes & T. Edison	England & U.S.A.	Independently invented the "carbon microphone." The "mic" converts sound (vibrations in the air) into an electrical signal.
1887	E. Berliner	Germany -U.S.A.	Invented the flat-disk phonograph. Disk records were longer playing and easier to store than cylinders.
1898	V. Poulsen	Denmark	Invented the magnetic wire recorder. Piano wire was pulled across an electromagnetic coil (or recording head) which magnetized each point along the wire in accordance with the incoming voice signal. Playback was through earphones.
1906	L. De Forest	U.S.A.	Invented the three-element "Audion" (triode) vacuum

			tube, the first practical amplification device to increase sound volume.
1928	F. Pfleumer	Germany -U.S.A.	Invented "magnetic tape": a thin coating of iron filings on a long, narrow strip of plastic.
1931	A. Blumlein	England	Invented "stereo sound." Used two microphones and two speakers. Record grooves had two sides, one for each sound track.
1933	E. Schuller	Germany	Developed the world's first tape recorder.
1956	Ampex Corp.	U.S.A.	Invented the "Video Tape Recorder." Live TV material could thereafter be saved for "delayed broadcast." In 1968 color video was added.
1966	J. Russel	U.S.A.	Invented the compact disc (CD), a digital data storage format. It replaced vinyl records.
1982	Sony	Japan	Launched the first CD player.

46. Motion Pictures …
Live-action moments become permanent, overriding time.

Year	Inventor	Country	Invention
1824	P. Roget	England	Argued that the eye holds an image for about one-tenth of a second. Therefore, to see an event in motion, at least 10 images per second must be seen.
1882	E.J. Marey	France	Made a "photo gun" capable of taking 12 consecutive frames in one second. With this tool he studied the motion of horses, birds, cats and a variety of animals.
1888	H. H. Reichenbach	U.S.A.	Working for George Eastman, Reichenbach invented celluloid (plastic) photographic roll film.

1892	T. Edison & W.K. Dickson	U.S.A.	Invented the first movie camera using celluloid film. The camera was powered by an electric motor which pulled the film along on its sprockets and holes, stopping to expose each picture frame, at least 10 frames per second as the film passed through the shutter.
1893	T. Edison & W.K. Dickson	U.S.A.	Invented the Kinetoscope, a peep show machine showing one continuous loop of film. Viewed individually through a glass lens in a cabinet.
1894	C.F. Jenkins & Lumiere Brothers	U.S.A. & France	Independently invented the world's first motion picture projector.

			Film can be viewed by more than one person. A sprocket wheel pulled the film before a light source which sent the picture through a lens and onto a screen for viewing.
1895	G. Méliès	France	Invented time-lapse photography, and hand-painted color film.
1916	D. Comstock & W. Wescott	U.S.A.	Invented "Technicolor." Camera lens exposed two strips of negative film simultaneously. Behind one was a red filter and behind the other a green filter. It was the most widely used color process in Hollywood from 1922 to 1952. Used in such films

			as Snow White and the Seven Dwarfs [1937], *The Wizard of Oz* (1939) and *Gone with the Wind* (1939).
1926	Warner Brothers	U.S.A.	*Don Juan*, was the first "talking movie." Dialog was recorded onto a wax record and synchronized with the projected film. The "sound-on-film" method, would become the standard for talking pictures in the 1930s.
1998	Texas Instruments	U.S.A.	Digital Light Processing (DLP) technology. Movies are shot, edited and shown digitally. Replaced movie "film."

47. Radio …

A person's voice is able to be in many places at one time.

Year	Inventor	Country	Invention
1819	H. Oersted	Denmark	Discovered that a wire carrying an electric current has a "magnetic field" around it.
1831	M. Faraday	England	Said that a magnetic field contains "lines of force."
1864	J. Maxwell	Scotland	Wrote math equations that predicted the existence of electromagnetic "waves" within a magnetic field.
1887	H. Hertz	Germany	Proved the existence of Maxwell's electromagnetic "waves." He used an electric coil to produce sparks. This spark "transmitter" produced the invisible

			"waves." Across the room was a "receiver," or a loop of wire with a gap. When sparks were produced with the "transmitter" coil, a spark would jump the gap in his "receiver" across the room. Airborne electromagnetic "waves" caused this event.
1894	G. Marconi	Italy	Invented wireless telegraphy. His goal was to use "Hertzian waves" to send telegraph messages without using wire lines (wireless telegraphy). In 1895 he succeeded in sending a wireless signal over one mile. In 1899 he sent a wireless

			message across the English Channel between England and France. And in 1901 he sent wireless signals between Cornwall in England and Newfoundland in North America, a distance of 2100 miles. These transmissions did not involve voice messages. Rather he used a spark-gap transmitter and a "coherer" to send and hear telegraph-like signals.
1898	N. Tesla	Croatia-U.S.A.	Operated a radio-controlled boat in a pool of water at Madison Square Garden in New York City. Crowds were

			stunned. He also was able to turn the boat's lights on and off with radio waves! Tesla is therefore thought to be the inventor of radio.
1900	R. Fessenden	Canada-U.S.A.	Was first to broadcast the human voice over the airways. In 1900 on Cobb Island on the Potomac River (50 miles from DC) he transmitted the human voice one mile between two 50-foot towers. He used something like a "crystal radio" to hear the voice message. He realized that because the human voice is always changing, a high-

			speed spark transmitter would be required. Such a high-speed transmitter would produce a "continuous wave" that could be "modulated" to match the changing wave form of the human voice. He used a carbon microphone to modulate this carrier-wave. This then, was "amplitude modulation," or AM radio. In 1901 he patented a high-speed AC dynamo to produce such a carrier wave.
1906	L. de Forest & R. von Lieben	U.S.A. Austria	Separately invented the 3-element "vacuum tube," or Audion. The vacuum tube

			was the first "amplification" device. It made possible "radio broadcasting."
1912	E. Armstrong	U.S.A.	Discovered the "Feedback Circuit" in vacuum tube amplification. By putting some of its output back into the tube over and over again, amplification increased hundreds of times. For the first time amplified signals were strong enough, so that "radio receivers" could use "loud speakers" instead of head phones.
1916	L. de Forest	U.S.A.	Radio station 2XG was an experimental station in New York City licensed to Lee De

			Forest. It had news and entertainment broadcasts on a regular schedule. It also used the first vacuum-tube transmitter. As vacuum tube "feedback" increased beyond a certain point, it would become a "continuous-wave" transmitter.
1920	KDKA	U.S.A.	KDKA in Pittsburgh received the first federal radio license and began broadcasting in November 1920. By 1922 there were 500 radio stations in the USA.
1928	CBS	U.S.A.	The Columbia Broadcasting System (CBS) was formed

1947	J. Bardeen, W. Brattain, & W. Shockley		Invented the "transistor." Tiny transistors replaced the big vacuum tubes of earlier radios. While vacuum tubes often burned out, transistors did not. Radios also became much smaller in size.
1954	Texas Instrument	U.S.A.	The Regency TR-1 which used transistors was the world's first commercially produced "transistor radio." Transistors made the "pocket radio" possible.

48. Television …
A person can be present in different places at one time.

Year	Inventor	Country	Invention
1875	W. Crookes	England	Invented the "Crookes Tube" which is the basis for

			the "tube TV." It consisted of a glass tube with the air removed. Two metal electrodes were placed inside, one at each end: the cathode (-) and the anode (+). When high voltage was applied at the cathode end, "cathode rays" flew in straight lines across the tube to the (+) charged anode. In 1897, these rays were revealed to be "electrons" by J.J. Thompson at Cambridge.
1895	J. Perrin	France	Invented the "Perrin Tube." It was an air-free glass tube with an "electron gun" at one end, sandwiched between two magnetic plates, shooting electrons at a fluorescent screen on the other end of the tube.

			Its electron beam was controlled by the two magnets.
1897	F. Braun	Germany	Modified the Crookes tube by adding a phosphor-coated screen.
1929 - 1933	V. Zworykin	Russia – U.S.A.	Invented the first electronic TV camera called the "Iconoscope." It collected an image on a charged plate containing photo-sensitive silver granules. An electron-beam then periodically swept across the plate, effectively scanning the stored image. He also invented the first electronic TV receiver, the "Kinescope," having only 60 lines of definition.
1946	A. Rose & P. Weimer	U.S.A.	Invented the "Image Orthicon" TV camera at RCA. It had a

			viewfinder; was only 15 inches long and 3 inches wide. In 1946 it was used to "tele-vise" the Joe Louis-Billy Conn heavy-weight fight at Yan-kee Stadium.
1950	Weimer, Forgue & R. Goodrich	U.S.A.	Invented the "Vidi-con" camera with 483 scan lines of resolu-tion. It was only 6 inches long and 1 inch wide.
1954	RCA	U.S.A.	The first color TVs were sold to the pub-lic.
1956	Ampex Corp.	U.S.A.	Invented the TV video recorder. The VRX-1000 allowed 90 minutes of record-ing. Delayed TV broadcasts became possible.
1964	G. Heilmier	U.S.A.	Invented the "Liquid Crystal Display" (LCD). He discov-ered several electric-optic effects while

			studying liquid crystals.
1977	J. P. Mitchell	U.S.A.	Developed the first LED (Light Emitting Diode) flat-panel TV screen. (Based upon work done in 1962 by Nick Holonyak at General Electric.)
1983	Casio	Japan	The Casio TV-10 was the earliest LCD (Liquid Crystal Display). television. Screen size: 30 inches. By 2005: 82 inches. LCDs are light weight, compact, thin, reliable and have better resolution than the "tube TV." LCDs have a thin layer of liquid crystal, sandwiched between two electrical conducting plates.

49. The Computer …
extending the human mind

Year	Inventor	Country	Invention
1943	T. Flowers	England	Designed and built the "Colossus," the world's first programmable, electronic, digital computer. It was used to unlock encrypted German messages during WWII. It had 1,600 vacuum tube circuits and a paper-tape input. It used Boolean algebra to do logical operations on its data.
1945	J. Mauchly & J. P. Eckert	U.S.A.	Built the "Eniac" (Electronic Numerical Integrator and Computer) at the University of Pennsylvania. It was similar to Colossus, but much faster. It could add or subtract 5,000 times a second. It was

			room size, weighed 30 tons, and had 18,000 vacuum tubes. It had a punch card input and output. However, it could do only one task at a time, since it had no operating system.
1947	J. Bardeen, W. Brattain, & W. Shockley	U.S.A.	Invented the "transistor" at Bell Labs. It was about an inch square, had 3 layers of the semiconductor silicon, and served the same function as a 3-element vacuum tube. However, it was smaller, faster, cheaper, produced little heat, used less electricity, and rarely needed replacing.
1957	I.B.M.	U.S.A.	The IBM 7090 was one of the first all-transistorized computers in the world. It had 50,000 transistors. It was used in

			the development of the Saturn 5 moon rocket in the 1960s.
1959	J. Kilby & R. Noyce	U.S.A.	Invented the "integrated circuit." Transistors had a problem: They needed to be soldered together. The more complex the circuits, the more numerous the soldered connections. The Noyce idea was to "print the circuits." Soldering would then be little needed.
1966	MIT	U.S.A.	The AGC (Apollo Guidance Computer) was developed at MIT in the early 1960s. It first flew in 1966. It was one of the first integrated circuit computers, designed to help astronauts land on the moon.

1967	D. Engelbart	U.S.A.	Invented the computer "mouse" at the Stanford Research Institute (in California) as a pointer or graphics interface.
1971	F. Faggin	Italy-U.S.A.	Built the world's first single chip "microprocessor" at Intel Corporation. It was the Intel 4004 chip, with a "processor," memory. and input-output controls. It integrated 2,300 transistors on one chip. What filled a room in the 1940s, would now sit in the palm of one's hand. (The modern chip presently has over a billion transistors on it, in an area the size of one's thumb nail.)
1974	H.E. Roberts	U.S.A.	Invented the first personal computer, the Altair 8800, using the Intel 8080 chip.

1975	Bill Gates & Paul Allen	U.S.A.	Bill Gates and Paul Allen wrote BASIC compiler for the Altair 8800 and formed Microsoft in Albuquerque, New Mexico.
1977	S. Wozniak & Steve Jobs	U.S.A.	Built the Apple II, an early personal computer. It had integrated circuits, a keyboard and monitor, color graphics, and a 4,100-character memory. Programs and data were stored on a cassette recorder. Wozniak (26) and Steve Jobs (21) sold 2 million Apple II computers!
1980	Y. Yokozawa	Japan	Invented the first laptop, or "notebook" computer: the Epson HX-20. It was introduced by Seiko in 1982.
1982	J. Russel	U.S.A.	Invented the "CD" in 1966.

			At first it was used for sound storage. Then, in 1982 it was adopted for use in computer storage: the CD-Rom.
1984	S. Jobs	U.S.A.	Apple started selling the Macintosh computer. It was the first computer with a graphics user interface and a "mouse."
1985	Microsoft	U.S.A.	Introduced "Windows," a graphics operating system for PCs.
1999	D. Moran, O. Ogdan, & A. Ban	Israel	Invented the USB "flash drive" (also known as the "thumb drive" or "memory stick.") IBM began selling the first flash drives in 2000.

50. Satellites ...

sending radio and TV persons to all parts of the Earth

Year	Inventor	Country	Invention
1954	D. Chapin & C. Fuller & G. Pearson	U.S.A.	Built the first Silicon Photovoltaic Cell (the Solar Cell) to produce electricity from sunlight. At first 4% efficient; later 11% efficient. Satellites are primarily powered by solar cells.
1957	M. Tikhonravov	U.S.S.R.	Sputnik 1, the world's first satellite, was put into orbit. It had a radio transmitter but could not send data from one point to another on Earth.
1958	K. M. Smith	U.S.A.	Project SCORE (Signal Communications by Orbiting Relay Equipment) was the world's first communications satellite. SCORE was the first

			communications re-lay system in space. It broadcast a Christmas message via shortwave radio from President Eisenhower through an on-board tape recorder.
1960	WDL	U.S.A.	Courier 1B was the world's first satellite to send messages to ground stations as it traveled. Courier was built by California–based Western Development Labs (WDL). It used solar cells to recharge an onboard battery.
1962	John Pierce & R. Kompfner	U.S.A.	Telstar 1 relayed the first television pictures and telephone calls from space. It did the first TV transmission from Europe to North America in July, 1962. One hundred

			million people watched!
1962	NASA & RCA	U.S.A.	Relay 1 was the first satellite to broadcast live television from the United States to Japan.
1964	Hughes Aircraft	U.S.A.	Syncom 1, 2 and 3 were the first geosynchronous communications satellites. Syncom 3 was used to telecast the 1964 Summer Olympics live from Tokyo to the United States.
1965	Hughes Aircraft	U.S.A.	Intelsat provided live TV coverage of the splash-down of the Gemini 6 spacecraft in 1965. By 2011 Intelsat had a fleet of 52 communication satellites above the earth.
1976	Taylor Howard	U.S.A.	Built the first home satellite TV dish. At

			first 20 feet in diameter; 6 feet by 1980 and 4 feet by 1990.
1993	Easton, Getting & Parkinson	U.S.A.	The first experimental GPS satellite was launched in 1978. By December 1993, GPS achieved initial operations with a full group of its own satellites. GPS has revolutionized navigation both at home and at sea. Those credited with inventing it: Roger L. Easton of the Navel Research Lab, Ivan A. Getting of Aerospace Corp, and Bradford Parkinson of the Applied Physics Lab.

51. The Internet …

A person can be seen and heard anywhere on earth as he speaks, which is an image of God, who is everywhere present!

Year	Inventor	Country	Invention
1961	L. Kleinrock (MIT)	U.S.A.	Published a paper on "Packet Switching Theory." The paper was titled: "Information Flow in Large Communication Nets." In this theory, data is broken up into "packets" that can be exchanged throughout a network of users.
1965	L. Roberts (MIT)	U.S.A.	Connected a TX-2 computer in Boston to a Q-32 computer in California. They were connected by a telephone line, forming the first wide-area "computer network."

1966	L. Roberts (MIT)	U.S.A.	Went to ARPA, the "Advanced Research Projects Agency," an arm of the U.S. Department of Defense, to develop the concept of a computer network.
1968	ARPA	U.S.A.	ARPA funded the formation of the "Advanced Research Projects Agency Network" (Arpa-net). Its goal was to link computers at a number of universities by way of telephone lines. It was an early "packet switching" network.
1968	BBN	U.S.A.	ARPA contracted Bolt, Beranek & Newman (BBN) to build the first "router," or "Interface Message Processor" (IMP). It was a switching platform

			that connected participating networks to the ARPA-NET.
1969	L. Kleinrock	U.S.A.	Sent the "first Internet message" from UCLA. The first internet "nodes" were at UCLA and Stanford University. By 1975 there were 57 such nodes. By 1981 there were 213.
1971	R. Tomlinson	U.S.A.	Invented "email" for the ARPA-NET system, the precursor of the Internet.
1973	B. Kahn & V. Cerf	U.S.A,	Wrote "Transmission Control Protocol" and "Internet Protocol" (TCP/IP). This is a model for how data is to be transmitted between multiple networks. "Open architecture networking" was their key idea.
1978	W. Christensen &	U.S.A.	Developed the first public dial-up BBS

	R. Suess		(bulletin board system), a computer server that allows users to connect to the system using a terminal program. Once logged in, the user can perform various functions, such as uploading and downloading software and data.
1983	P. Mocka-petris & J. Postel	U.S.A.	Invented the "Domain Name System" (DNS). By 1987, the number of Internet "hosts" exceeded 20,000.
1989	T. Berners-Lee	U.K.	Invented the "World Wide Web," and built the first Web browser. (The WWW is not to be confused with the Internet. The WWW is a means of accessing web sites on the Internet.)

1994	J. Yang & D. Filo	U.S.A.	Created "Yahoo," a search engine with a directory of websites.
1995	J. Bezos	U.S.A.	"Amazon" goes on-line.
1996	Page & Brin	U.S.A.	"Google" was begun by Larry Page and Sergey Brin at Stanford University. Google is a search engine and on-line library.
1997	Hastings	U.S.A.	"Netflix" was founded by Reed Hastings and Marc Randolph.
2004	M. Zuckerberg	U.S.A.	Created Facebook, a social network website. By 2006 it had 400 million active users.

The Internet has become the global means of communication. In 1993, the Internet involved only about 1% of the world's communication. By 2017, 51% of the world's population had Internet access.

52. The Smartphone …
One can see and talk with almost anyone on earth.
Thus, space and time bounded-ness
is disappearing – a sign of Heaven.

Year	Inventor	Country	Invention
1973	M. Cooper (Motorola)	U.S.A.	Invented the first hand held mobile phone. It was 10 inches long and 2.5 pounds.
1983	Motorola	U.S.A.	The Dynatac 8000x was the first commercially sold mobile phone at a cost of $4000 each. It was 13 inches high and weighed 2 pounds. It took 10 hours to recharge its battery.
1992	F. J. Canova (IBM)	U.S.A.	Developed the "Simon Personal Communicator." It is credited as the world's first handheld, touch screen Smartphone. It was able to send and receive email,

			faxes, and regular phone calls. It was 8 x 2.5 x 1.5 inches and weighed about 1 pound. It had a touch screen keyboard, address book, calendar, electronic notepad, and internet access. It was released under the name "Simon" in 1994. Between 1994-1995, fifty thousand "Simons" were sold.
1996	Nokia	Finland	The "Nokia 9000" was 7 x 3 x 1.5 inches, and weighed a little less than a pound. It had a punch keyboard and screen. It offered email, fax, and normal phone uses. It had Internet access but no camera.
1999	M. Lazardis	Canada	Developed the "Blackberry" Smartphone. It had a punch keyboard and

			a black-white screen. Was 2.5 x 3.5 x 1 inches in size. Had email messaging and Internet browsing (but no camera). By 2013 the Blackberry, with updates, had sold 85 million units worldwide.
2003	J. Friis & N. Zennstrom	Denmark Sweden	Invented Skype. Skype allows speakers to be seen on their smartphone or computer screens while talking with someone across the planet.
2007	Steve Jobs (Apple Computer)	U.S.A.	Developed the "I Phone." It was half the size of IBM "Simon": 4.5 x 2.5 x .5 inches and had a multi-use touch screen, plus a 2-megapixel camera. It had the iTunes store and the "Safari" web browser. In the first

			year, 2007-2008, Apple sold 6 million I Phones.
2008	Apple App Store	U.S.A.	The "Apple App Store" was put online with 500 new apps to choose from. The word "app" means "application." An app is a piece of software. There are apps for social networks, games, travel, banking, health, news, and many other topics. The Apple App Store adds about 20,000 new apps each month. In one year, 2014, 138 million apps were downloaded.

By 2008, most smartphones had GPS (the Global Positioning System). GPS makes it possible to find a location anywhere on earth! By 2012, studies indicated that there were over 1 billion smartphones being used worldwide.

The smartphone is a sign of life in Heaven. It gives everyone power over space and time. Google (1996) gives each person the resources of a library and thereby mimics the power of the mind in Heaven. "Wikipedia" (2001) provides the power of an encyclopedia, deepening one's overall knowledge base. Google's "visual globe" (2005) and "street-view" (2007) puts each person in touch with various of God's people around the world. These sites are pointers to life in Heaven.

FACTORIES
CREATING A "NEW WORLD"

53. The New World

Factories are the places where new things are made. They are the places where a new world is being fashioned. A factory can be housed in one room or on a city block filled with individual buildings. Factories are the places where new ideas get expressed in concrete ways. Factories are signs of Heaven. They make the invisible appear before us. They demonstrate God's creative-powers in that they imitate his initial creation of the world from nothing.

Factories are a powerful sign of God's continuing Presence. Their collective work is a sure sign of what God has prepared for humans beyond our powers of sight, on the other side of this life. Factories, viewed with the eyes of faith, are a sign of Heaven – the Kingdom of God.

54. Iron into Steel …
a sign of resurrection

Iron is the sixth most abundant element in the universe, and the fourth most abundant element in the earth's crust, about 5%. Iron is a gift from God to the people of Earth. It is the most often used metal in making machines, vehicles, bridges, and building structures.

Pure iron is not found in nature because it readily unites with oxygen in the air, forming iron oxide (what is called "rust"). A common source of iron is hematite, one of the iron oxides.

To remove iron from iron ore requires that it be melted in a very large furnace, where oxygen and other impurities are removed. Iron, however, in its pure form is not very strong. Pure iron is too soft and reactive to be of much use in this world. In fact, it is possible, with some difficulty, to cut through a chunk of pure iron using only a knife!

The iron that one sees in the skeletons of buildings, in ships and automobiles, is primarily a mixture of iron and carbon, what is called "carbon steel." The amount of carbon needed to turn iron into steel varies from about .05% to 2.0%. On average, most of the steel one sees in large buildings is roughly .5% carbon. That means that for every 100 pounds of iron, about ½ pound of carbon is necessary, to turn soft iron into steel. Carbon atoms in the steel hide between the iron atoms to strengthen or "steel" them. The amount of strengthening that occurs in this process is truly amazing. Steel is 1000 times stronger than pure iron!

This transition from iron into steel can be viewed as a model for the transition of a person on Earth into that same person in Heaven – where one's life will be 1000 times more satisfying or happy than here on this planet. In other words, one's body here on Earth resembles the soft and reactive nature of iron, whereas in Heaven one's body will be stronger and more useful. Notice, for example, how the Bible speaks of life in Heaven.

"God himself will be with them. He will wipe away all tears from their eyes, and death will be no more; nor will there be mourning, crying, or pain …"

Revelation 21:3-4.

What follows are a few more insights into what risen life in Heaven will be like. This examination will primarily look at one day in the life of Jesus, that of the first Easter Sunday. In the events of that first Easter, one can see Jesus in his risen body, a preview of what our own expansive life in Heaven may look like.

The Risen Jesus

In the New Testament, it is recorded that Jesus raised many people from death back to ordinary earthly life. These people would die a second time. However, resurrection is life beyond nature; resurrection is supernatural life. For example, when Jesus was on the Cross, he made the following statement to the repentant thief: "*This day* you will be with me in Paradise" (Luke 23:43). That was an amazing statement. What it means is that between Jesus's burial on Good Friday and his resurrection on Easter Sunday, Jesus would be in Heaven with the Good Thief! In the NT, "Paradise" is another word for Heaven (see Revelation 2:7 and 2 Corinthians 12:4).

When Easter Sunday did arrive, certain women went to the burial site of Jesus, to embalm his dead body. They brought spices to do this work (Luke 24:2). They did not expect to be told that Jesus had risen. However, Jesus was not there! An angel informed the women that Jesus had already risen, going through the walls of his tomb (Matthew 28:2-6).

In the middle of the afternoon on that same day, Easter Sunday, Jesus joined two of his disciples on a walk from Jerusalem to the town of Emmaus about seven miles away (Luke 24:13). Jesus was in his risen body on this walk, but his two disciples did not recognize him! Of this, Luke said: "their eyes were being held" (Luke 24:16), but Mark offered another explanation: "He (Jesus) was shown in *another form*" (Mark 16:12). In other words, Mark said that Jesus's risen body was in some sense different. It was Jesus to be sure, but transformed.

Hours later, toward evening, these two men invited Jesus to stay overnight with them. Jesus agreed, and at some point in the early evening, broke bread with them. In this action of "the breaking of the bread" (Acts 2:42), another insight into the nature of risen life would be seen. Jesus took bread, blessed it, and gave a piece to each of the two men. At this point Scripture says: "their eyes were completely opened" (they recognized who he was), and then it continues on: "he disappeared from them" (Luke 24:31). In other words, he instantaneously left them! Why so? Because he was now present in the transformed Bread, equally present as he had been in his risen body minutes before! How do we know this?

At the Last Supper, three days earlier he had promised that when using his words of consecration, such bread would become his Body: "having taken bread and given it his blessing, he broke it, and gave it to them (the Twelve Apostles), saying: 'This is my Body ... do this (action) for my remembrance" (Luke 22:19). Thus, at this table in Emmaus, Jesus demonstrated two things: (1) The Eucharist will open one's eyes to see his real Presence (Luke 24:35). And (2) risen life is not subject to the laws of nature, or specifically, to the rules of space bounded-ness.

After leaving these two men, Jesus appeared later that same evening to his Twelve Apostles back in Jerusalem. In that event, Jesus came through closed doors, possibly through the walls of the site (John 20:19 and Luke 24:36). His own apostles were "terrified" by this happening, thinking for a second or two that Jesus was a "spirit." Jesus then said to them: "Why are you troubled?" (Luke 24: 38). He then proceeded to show them his hands and his feet with the nail holes, to assure them that he was in a real body, even though this risen body possessed special powers. And finally, to make the realness of his body even clearer, he ate some "grilled fish" that they had prepared (Luke 24: 42-43).

Thus, it can be seen that the resurrection of Jesus, involved a real body that had been transformed – like iron into steel. Jesus's body is the living model for the resurrection of each person on earth. The risen body of each person will be real like that of Jesus but freed from the laws of nature that surround us. Risen people are, in a sense, like a pure form of iron here on Earth, but destined to become steeled for life in higher places.

55. Aluminum ...
creating new metals

God has created our solar system using 92 naturally occurring elements. Of these 92 different atoms, about 68 are metals. The most abundant metal in the crust of the Earth is aluminum, about 8%. That is another of God's gifts to people on earth. However, aluminum does not occur in a pure form on the Earth. Aluminum is always tightly held by one or more additional elements.

The ore in which aluminum is found is called bauxite. This ore is found in such places as Australia (23%), China, Brazil, Guinea, and in "Bauxite, Arkansas." Bauxite is not a "mineral" in which just two or three elements are bound together. Bauxite is a "rock" in which several minerals exist side by side. Roughly 5300 different kinds of "minerals" have been identified in nature.

In 1825, Hans Christian Orsted of Denmark discovered the metal aluminum. He isolated it from one of its compounds, "Aluminum chloride." Later, small amounts of aluminum were obtained by heating aluminum ore with sodium in a vacuum. These, however, were small amounts. Then, in 1886 Charles Hall (USA) and Paul Heroust (France) separately discovered a process to extract large amounts of aluminum from its bauxite ore. They reduced bauxite to aluminum oxide (alumina), then changed this white powder into a liquid by adding cryolite, and lastly removed the oxygen by sending an electric current through this liquid. (AlO_2 + electricity → aluminum + oxygen).

In 1888, Hall opened the first large scale aluminum plant in Pittsburgh, Pennsylvania. Hall was 25 years old! It later became the Alcoa Corporation.

The process of discovering such a metal and then using it creatively has three distinct steps: (1) The long term action of God in giving humankind aluminum ore. (2) The various individuals who discovered God's gift in the crust of the Earth and learned how to extract it, and (3) Those who mix aluminum with other metals to produce newly created, designer metals, to accomplish various specific jobs here on Earth. These achievements make mankind a co-creator with God and mirror the fact that we have been created in God's image [Genesis 1:26].

The third of these three steps deserves a closer look. The "Aluminum Association" has stated that there are currently 530 recipes to make different kinds of aluminum alloys. An alloy is made by adding different chemical elements to pure aluminum metal. Usually, these different elements are other kinds of metal atoms, such as copper, manganese, iron, or zinc.

Pure aluminum metal by itself is weak. However, if alloyed to other metals it becomes tough, and in a few cases as strong as steel.

Aluminum is a very useful metal. For example, it is only about a third as heavy as iron, has an attractive shiny-white luster, and is a good conductor of electricity and heat. However, these attractive characteristics can be improved, especially in the area of strength.

What follows are a few examples of how mankind's creativity is fashioning a "new world" by inventing new aluminum alloys.

1	Pure Aluminum	99.9% pure	Used to make "aluminum foil" for the kitchen. Also, for sheet metal applications, such as clothes washers and dryer bodies.
2	Aluminum 2024	4.5% copper 1.5% magnesium .6% manganese	A widely used high strength aluminum alloy. Used in aircraft bodies and wings, truck wheels, auto parts, screws and rivets, recreational equipment, etc.
3	Aluminum 3003	2.5% manganese	A general purpose alloy. Used to make beverage (or "pop") cans, cooking pots and pans, pressure vessels, hardware, architectural components, aircraft cowlings, etc.
4	Aluminum 5052	2.5% magnesium .25% chromium	Stronger than the 3003 alloy. Used in fuel tanks, marine applications, hydraulic tubes, boat hulls, trucks, etc.

| 5 | Alumi-num 6061 | 1.0% magnesium .6% silicon .25% copper .25% chromium | Widely used alloy. In aerospace components, truck frames, machine parts, valves, camera bodies, fishing reels, electronic parts, railings, etc. |
| 6 | Alumi-num 7075 | 5.6% zinc 2.5% magnesium 1.6% copper | A very high strength alloy. Used in aircraft structures, recreational equipment, the Bradley Fighting Vehicle (for the army), etc. |

One concluding point. Aluminum is but one metal in the Periodic Table of chemical elements. There are 67 other kinds of metal with a history similar to that of aluminum. The alloys of many other metals have also contributed to the building-up of this "new world" – and contribute to a future image of mankind in Heaven. This "new world," being fashioned on this planet, is, however, but a temporary reality. We are leaving this planet. The "world" that we know so well here will be "history for us" in a few years. Nonetheless, God has revealed our true home, which is spoken of as the "Kingdom of God" in the Bible – or "Heaven" in the language of ordinary folks. (Mark 1:15 & Luke 10:9).

56. Plastics …
many new things, as a sign of Heaven

Plastics are made of long chains of small molecules (monomers) bound together into one very large molecule called a "polymer." God has been knitting polymers together, in nature since the beginnings of life on Earth. Every living plant or animal contains these long molecular chains. The cell walls in plants are made from the polymer "cellulose." The proteins in our skin and muscles are made of polymers; as is the spiraling structure containing our body's basic building plan, known as DNA.

Most polymers are based upon the atom carbon. Carbon is the backbone of the various polymer molecules. Carbon is strong and stable and has four important sites where other atoms can join it. Side chains, containing hydrogen, oxygen, nitrogen, and other atoms determine what a specific polymer will do inside the body.

Polymers can be natural or synthetic. Synthetic polymers do not exist in nature. They are constructed in labs by chemists who put atoms together in new ways. This section will first examine polymers found in nature and then look at polymers created in laboratories. Synthetic plastics, that are everywhere present in the modern world, is the specific topic of this section.

Year	Inventor	Country	Invention
1838	A. Payen	France	Discovered "cellulose." Isolated it from plant material and determined its chemical formula.
1846	F.J. Otto	Germany	Invented "gun cotton" (nitrocellulose). He used nitric acid to convert cellulose into nitrocellulose and water.
1855	A. Parkes	England	Invented the first plastic which he called "Parkesine." He discovered that mixing nitrocellulose and camphor produced a mass that could be molded when hot, and then became hard when cold. In 1862 he showcased Parkesine at the International Exhibition in London.
1869 -1872	J. & I. Hyatt	U.S.A.	Hyatt plastic was essentially Parkesine

			with added heat and pressure. Isaiah Hyatt named it "Celluloid." The Hyatt brothers also invented an "injection molding machine" which had a plunger to inject hot plastic through a cylinder into a two-part mold. It was used to make billiard-balls, false teeth, and piano keys.
1891	C. F. Cross & E. Bevan	England	Invented "Rayon," or artificial silk. It was made from wood pulp, a natural source of cellulose. Used in dresses, slacks, jackets, work clothes, ties; bedspreads, blankets, sheets, curtains, and surgical products.
1893	A. Trillat	France	Developed "Galalith," a type of plastic still used to

			make dress buttons. He immersed casein from milk in formaldehyde to create this product.
1907	L. H. Baekeland	Belgium-U.S.A.	Created the first fully synthetic thermoset plastic called "Bakelite." He mixed phenol, a waste product of coal, and formaldehyde. Earlier plastics had limited usefulness because of their tendency to soften when heated. Bakelite remained hard even in heat, and therefore was used to make many things: radios, phonographs, combs, jewelry, cars, airplanes, etc.
1909	F. Hofmann	Germany	Made the first synthetic rubber, Isoprene. By 1910 car

			tires were made with it.
1912	J. Branden-berger	Switzer-land	Invented "Cello-phane," a light, non-reactive, plastic wrap. It became very successful after W. Church at DuPont made it wa-ter-proof with a ni-trocellulose lacquer in 1927. Used today to wrap candies.
1920	H. Staud-inger	Germany	Proved that cellu-lose, starch, pro-teins, and even rub-ber are made of short repeating mol-ecules linked to-gether like paper clips hooked to one another. These mac-romolecules he called "polymers." He received the 1953 Nobel Prize in Chemistry.
1925	R. G. Drew (3M)	U.S.A.	Invented "Scotch Tape," the first

			transparent, sticky tape.
1926	W. Semon	U.S.A.	E. Baumann of Germany accidently synthesized "Polyvinyl chloride" (PVC) in 1872. Waldo Semon of B.F. Goodrich developed a method in 1926 to plasticize PVC by blending it with additives. It is presently the world's third most used plastic (C_2H_3Cl). Used in plumbing pipe, doors and windows, cable insulation, imitation leather, packaging, inflatable boats, etc.
1930	A. Collins & W. Carothers	U.S.A.	Invented "Neoprene," presently the most widely used synthetic rubber.
1931	I.G. Farben Company	Germany	E. Simon (Germany) began the

			history of "Polystyrene" in 1839 when he accidently discovered a monomer named styrene. This was an oily substance. In the 1920s it was learned that this monomer would grow into a polymer when adequately heated. In 1931 the company I.G. Farben began manufacturing "Polystyrene," a high impact plastic used in plastic spoons, knives, and forks, CD and DVD cases, bottles, trays, tumblers, and yogurt containers.
1932	O. Rohm & W. Bauer	Germany	Invented "Plexiglas" (also known as safety glass, or Lycite). They put a plastic sheet (methyle methacrylate)

			between two layers of glass. Upon breaking, the glass would not shatter. Used in windshields for autos.
1933	E. Fawcett & R. Gibson	England	"Polyethylene" was discovered by accident in 1898. However, a practical method to make Polyethylene was not discovered until 1933 in England by Fawcett and Gibson. This is the most widely used plastic today. There is probably a polyethylene object near everyone on this planet. It is used in beverage bottles, plastic grocery bags, the lining in milk cartons, the bag inside a cereal box, milk and detergent

			jugs, water pipes, children's toys, etc.
1934	O. Rohm & O. Haas	Germany-U.S.A.	Invented the first usable acrylic resin. Between 1946-1949 L. Bocour and S. Golden developed acrylic paints.
1935	W. Carothers (DuPont)	U.S.A.	Invented "Nylon" fiber. It is light weight, very strong, and much less expensive than silk. Used for women's stockings. During WW II it was used in parachutes, tents, ropes and ponchos. In 1938 the nylon tooth brush came into use.
1937	O. Bayer	Germany	Discovered "Polyurethane." Used in coatings, car seats, shoe soles, foam insulation, etc.
1938	R. Plunkett (DuPont)	U.S.A.	Discovered "Teflon," a white, waxy material. Found it to

			be resistant to heat, chemically inert, with very low surface friction (nothing would stick to it). Used in nonstick pots and pans for the kitchen.
1939	E. Rochow	U.S.A.	Produced the first "silicone plastic," methyl silicone. Known for water repellency, thermal stability, and inertness. It Is widely used for lubricants, protective coatings, paints, electrical insulation, and prosthetic replacements for body parts.
1941	R. McIntire (Dow)	U.S.A.	Invented "Styrofoam." Used in coffee cups, and as a cushioning material in packaging.
1941	J. R. Whinfield & J.T. Dickson	England	Produced the first "Polyester" fiber. Much of the early

			research on this fiber had been done by Wallace Carothers at DuPont in 1930. Polyester clothing is resistant to stretching and shrinking; is easy to clean, and resists wrinkling and tears. Today its fibers are crimped and bulked to make polyester very soft.
1941	DuPont	U.S.A.	Produced the first "Acrylic Fiber," under the trade name "Orlon."
1942	H. Coover	U.S.A.	Invented "Super Glue" (methyl cyanoacrylate) at Eastman Kodak.
1951	J.P. Hogan & R. Banks	U.S.A.	Invented "Polypropylene"(PP) and "High Density Polyethylene" (HDPE).
1955	G. de Mestral	Switzerland	Invented "Velcro, the "hook and loop" fastener.

1959	J. Shivers (DuPont)	U.S.A.	Invented the "Spandex Fiber," known for its elasticity. Is stronger than natural rubber.
1965	Stephanie Kowalik	U.S.A.	Invented the "Kevlar Fiber" at DuPont. Kevlar is extremely strong, lightweight, and heat resistant. Used in bullet-proof vests and to make super-strong rope.
2009	Boeing	U.S.A.	The Boeing 787 airliner is 50% plastic, including all of its skin.
2016	M. Kanan (Stanford)	U.S.A.	Developed a novel method to create plastic from carbon dioxide and inedible plant material - rather than from petroleum.

Plastics have freed humans from many of the restraints of the natural world: from its limited supplies and from its limitations of form. Today, ordinary people benefit from plastics, a material that even kings of an earlier age could not have imagined.

All of this surely points to an even greater future — in Heaven!

57. Food …
"Look at the birds …
your heavenly Father feeds them …"

Mt 6:26

These words of Jesus are not just about the birds. These words are also about us. Most of the plants and animals we eat today were on this planet before we humans arrived. We have learned how to work with God; how to breed larger and stronger varieties. But that is all. We basically eat the same foods as we have always eaten, what were provided for us early on.

(1) Wheat

Wild wheat is a three-foot tall grass plant that originated in the Israel and Iran area of the Middle East. About 8000 BC, the semi-wild "Einkorn" wheat began to be grown there for food. It had 14 chromosomes and small seeds. The seeds were ground and mixed with water to form a "gruel." This particular kind of wheat has been found at the archaeological sites of Cayonu and Cafer Hoyuk in southern Turkey.

Another kind of wild wheat that came to be grown about the same time, in the same area, is known as "Emmer." Emmer wheat had 28 chromosomes and larger seeds. It has been found in the ancient sites of Beidha and Jericho. It was also widely grown in ancient Egypt.

A third kind of early wheat was bread wheat, or "Duram." It has 42 chromosomes and the largest head of seeds. It is able to grow in colder middle latitude climates and has been found in ancient sites dating from 6,000 to 4,000 BC.

(2) Rice

Rice is a tropical five-foot tall grass plant that grows wild in SE Asia. People began growing rice for its seeds in southern China about 6000 BC. By 2,500 BC rice was being grown in the Ganges Valley of northern India. Rice farming reached Greece about 300 BC. By 800 AD, people in East Africa were growing rice, and, in the 1600s AD, rice also came to North America with its English settlers.

(3) Corn (Maize)

Maize or corn is a plant which originated in the hills around Mexico City. Today, it is the largest food crop on the planet, with wheat and rice being second and third. Wild corn began its long history as "teosinte grass." The original "ear" of corn was only 3 inches long and had 8 seeds. The "ear of corn" looked much like the head of an ancient wheat plant.

In 1953, corn pollen was found under the Fine Arts Building in Mexico City. This pollen was shown to be 60,000 years old. At Bat Cave in New Mexico, some 700 "popcorn" cobs with kernels were found dating to about 5,000 BC. These cobs were only 1 to 2 inches long and had up to 50 seeds on each example. More recently, at Cox Cat Cave in Tehuacan Valley, Mexico, some 6-inch corn cobs were found holding up to 100 seeds each. These larger cobs belonged to the Spanish period, around 1519 AD. Finally, in the last 200 years of controlled cross breeding the size of the typical ear of corn has doubled to about 12 inches.

(4) White Potato

The potato is among the top five food crops on earth. It is a very important food source in colder climates. Potatoes originated in the Andes Mountains of Peru and Bolivia in South America – before people were on earth. The potato is a "tuber" attached to the plant's underground stem. The potato was taken to Spain in 1533 by the Catholic priest, Father A. Cardan. By 1600, potatoes were grown as far away as India and Japan, and by 1700 in China. "French Fries" became popular about 1750 in Europe, and "potato chips" were invented in 1853 by George Crum, a native American cook in Saratoga, New York. A customer asked Crum for some very thin fried potatoes. Crum gave him "potato chips."

(5) Tomato

The wild tomato plant originated in the mountains of Ecuador, Peru, Bolivia, and Chile. Tomatoes were grown by the Aztecs as early as 700 AD. The Aztecs called it the "tamotle" and were growing it in 1519 when the Spanish first arrived. An early recipe for "tomato catsup" appeared in a 1792 cookbook called the "New Art of Cooking" by the USA author Richard Brigg. In 1876, Henry Heinz began manufacturing "Tomato Catsup" in Pittsburgh, Pennsylvania.

(6) Honey

Honey was the world's first sweetener. Cave people collected honey. A 7,000 BC rock painting inside Arana Cave near Valencia, Spain, shows a man high in a tree collecting honey from a hive with bees circling. There is also a 2400 BC wall painting in the tomb of Rekhmire in Luxor, Egypt. This painting shows workers using smoke to calm bees while they are collecting honeycombs.

(7) Chocolate

Chocolate contains "phenylethylamine." The human brain also produces this chemical. It prevents depression and is also present when people feel loved. Cocoa trees originally grew wild in Central America. A typical cocoa tree contains about 25 pods and has about 20 cocoa beans in each pod. In 1519, the Aztec ruler, Montezuma, served Don H. Cortez this drink: ground cocoa beans mixed with ground corn in water, and flavored with vanilla. It was served with a spoon! In 1828, C. Van Houten in Holland found a way to remove the fat from cocoa beans. Cocoa beans – fat = cocoa powder …

cocoa powder + sugar + hot water = hot chocolate

Then, in 1876, H. Nestle in Switzerland invented the "milk chocolate" candy bar. This was followed in 1894 by Milton Hershey's milk chocolate bar in the USA.

(8) Livestock

Like plants, certain useful animals were found in the wild. These plant eating animals were mild enough in disposition to be brought into human contact. Such animals could be tamed and used to support human life.

The "mouflon," or wild sheep, were found in the mountains of Iran and were domesticated about 10,000 BC. Sheep were raised for their milk, wool, meat, and skin.

The wild goat lived in eastern Europe and into southern Asia. Genetic analysis has shown that the Bezoar Ibex of the Zagros mountains is the ancestor of all domestic goats. The remains of ancient goats have been found in Ganj Dareh in Iran.

The source of the domestic pig is the Eurasian wild boar (Sus scrofa). Archaeological evidence suggests that pigs were being managed in the Tigris Basin at Cafer Hoyuk by about 11,000 BC. Pigs were brought to North America by Hernando de Soto in 1539 and other Spanish explorers.

The source of the domestic cow was the wild "aurochs" of Europe and Asia. Cows were being farmed by 8,500 BC. There is a cylinder seal from the Iraq area showing a herd of cattle between 4,100 and 3000 BC. According to the UN survey there were 1.4 billion cattle worldwide in 2011.

Wild chickens, or Red Jungle fowl, originally lived in northern India and parts of east Asia. They were domesticated both in southern China and India about 7,000 BC.

The wild turkey originated in the forests of North America, from Mexico to the eastern half of the USA. The "meleagridinae" turkey is known from 23 million-year old fossils. Also, a turkey fossil from the late Miocene (about 6 million years ago) has been found in Westmoreland County, Virginia.

(9) Milk and Cheese

Cows, goats, and sheep turn grass into milk which humans can then consume. People cannot eat the tough grasses that cover many parts of the earth. Cows are especially good at changing grass into large amounts of milk. Cattle can feed on tough grass because they have four stomachs rather than just one to process it.

In early times, when ice was the only refrigerant, cheese making became a way of preserving milk. The essential step in cheese making is to take the water out of the milk. About 6 of every 7 parts of milk is water. Cheese is milk minus water, plus flavoring.

In Poland, 34 perforated pottery vessels or "cheese strainers" have been found that date to 6,000 BC.

Swiss cheese was brought to Rome by Julius Caesar's army in 58 BC. Roquefort cheese, with its blue mold, was invented in what today is southern France. It was first mentioned by the Roman writer Pliny who died in 79 AD. In 771 AD, Brie cheese was eaten at the court of Charlemagne. Parmesan cheese was first recorded by a Catholic priest, A. Salimbene, in the thirteenth century. Parmesan is a seasoning cheese and the cornerstone of Italian cooking. It was invented near Parma in Italy. Cheddar cheese was invented at Cheddar (a village) in Somerset County in England. It was first mentioned by the historian W. Camden in 1600. Lastly,

Camembert cheese was invented by Marie Harel at Camembert (a town) in Normandy, France in 1790.

(10) Ice Cream

Ice cream is made from cream (80%), sugar (15%), and flavorings (5%). It is chilled until semi-solid. The origin of ice cream is unknown. However, the Café Procope in Paris, France, was selling ice cream in 1670.

In 1874, Robert Green of Philadelphia (USA) invented the "Ice Cream Soda."

ice cream + syrup + soda water = ice cream soda

In 1881, the "Sundae" was invented by Ed Barnes of Two Rivers, Wisconsin (USA). A customer wanted a dish of ice cream topped only with the syrup. He did not want carbonated water as used in sodas.

The term "Milkshake" was first used in print in 1885. At first, this term referred only to an egg-nog like drink, but by 1900 chocolate, strawberry, and vanilla syrups were being added to refrigerated milk and ice cream in various ice cream shops. Then, in 1911, Hamilton Beach's electric drink mixer speeded up the process of making "shakes."

In 1904, at the St. Louis World's Fair, the "Ice Cream Cone" was invented when E. Hamwi (Syria-USA) put ice cream in a waffle, rolled up into a cone-shape.

Finally, in 1920 the "Ice Cream Bar," or "Eskimo Pie" was invented by Chris Nelson in Onawa, Iowa (USA).

(11) Sugar

"Sugar Cane" is a 10-15 feet tall grass plant. Its stalk is 2 inches wide and is filled with a sweet spongy material. Half of our sugar comes from this "cane sugar" plant. The other half comes from the "Sugar Beet" that looks much like a large white carrot that weighs about 2 pounds and is about $1/5^{th}$ sugar.

Sugar cane grows wild in New Guinea, which is north of Australia. New Guinea gets 60 inches of rain per year and is hot year-round.

The first sugar cane, as a crop, was grown in India near Calcutta. The people of India learned how to separate sugar cane juice by 3-5 boilings to a solid product. Sugar cane was brought to America by Columbus in 1493. Then, in 1590, O. de Serres in France noticed that: "The (white) beet root when boiled yields a juice similar to the syrup of (cane) sugar." By 1747, A. Marggraf in Germany had proved conclusively that cane sugar and beet sugar were the same product, namely sugar. Then, in 1795, E.de Bore in the USA produced the first "granulated sugar." Before 1795, sugar was sold in the form of a loaf or cone shaped product. Finally, in 1801 A. Achard in Germany built the first factory to produce beet sugar. This was followed in the USA by a sugar beet factory in Massachusetts in 1838.

(12) Breakfast Cereals

"Breakfast" means to "break one's fast," referring to the first meal after a night's sleep with no food.

In 1895, Charles Post of Battle Creek, Michigan, made "Grape Nuts" cereal. It consists of wheat and barley nubs (no grapes and no nuts!).

In 1898, John and William Kellogg of Battle Creek, Michigan, invented "Corn Flakes." They boiled corn seeds and then flattened them.

In 1901, the Quaker Oats Co. invented "Oat Meal." They steamed oat seeds and then rolled them flat.

In 1924, the General Mills Co. in Minneapolis, Minnesota, invented "Wheaties." They boiled wheat seeds and rolled them into flat flakes.

In 1937, General Mills also invented "Cheerios."

(13) Drying Food

Bacteria and molds need water to grow. Drying food kills them. Early peoples dried food in the sun, and then stored such crops as wheat, rice, beans, peas, and corn in dry pits or cool caves. Grapes were dried to become raisins. Plumes were dried to become prunes.

In ancient Egypt, pictures from about 2,000 BC show workers salting meat and then hanging it in the sun. While drying removed the water for germs to grow, salting killed them.

In 1920, milk was dried in 400-degree F. air, producing "powdered milk." This was then boxed and sold. In 1934 the Lipton Tea Co. introduced dried soups in packages.

(14) Refrigeration

Germs in food multiply 400 times slower when stored between 40-50 degrees F, rather than between 50-60 degrees F.

Early peoples often stored their food in caves (45-55 degrees F.) or 3 feet underground where the temperature is a constant year-round 55 degrees F. By 1600 AD, "Ice Boxes" were being used in kitchens in France. In winter, blocks of ice were cut from rivers and lakes and then stored in "ice houses" for use in the summer months. By 1856, Boston, Mass, was using 156,000 tons of ice each year to prevent food spoilage.

In 1755, W. Cullen in Scotland proved that liquid ether could turn water into ice by simply standing next to it. The liquid ether took heat out of the water and turned itself into a gas. Then, in 1834, J. Perkins of the USA built the world's first "refrigerator." It consisted of two parts: (1) Liquid ether that took heat out of water-based foods inside a box. And (2) a "com-press-or" unit outside the box, to "press" the gaseous ether back into its liquid form – thus completing one cycle.

In 1877, the first railroad "freezer car" was built for Gus Smith in Chicago. It carried fresh meat to various US cities. Bacteria cannot live in cold below 40 degrees F.

About 1913, refrigerators for home use began appearing. At first, the compressor unit was situated on top of the refrigerator box. Later, it was put inside near the floor.

In 1923, C. Birdseye of the USA began to "fast freeze" perishable foods like berries. In this way many summer-only foods could be eaten year-round.

About 1930, the "Freezer Truck" also began carrying frozen vegetables, fruits, and meats to the local food stores.

(15) Canning

Canning uses heat to kill germs by sterilization. After receiving very high heat, an air-tight seal is placed upon a glass jar or can containing food, to keep out any new bacteria or molds.

In 1804, Nicholas Appert in Paris, France, invented the canning method. Food was heated to near boiling. Then, a cork was pushed into a jar's opening to make an air-tight seal.

By 1825 the steel can, coated with tin (the "tin can"), had been invented for canning. The "tin can" was invented by T. Kensett in New York.

In 1825, canned salmon appeared on the market. In 1839, canned corn, 1847, canned tomatoes, 1863, canned peaches, 1873, canned soups by the Campbell Co., and in 1875, canned B&M baked beans.

(16) Cuisines

Various food traditions have developed over time among different cultures and peoples. There are, for example, French cooking, Italian cooking, Chinese cooking, Japanese cooking, and Mexican cooking. Other examples do exist.

Each of these different ways of preparing food is the result of countless inventions by forgotten individuals. One example (not forgotten) is chocolate chip cookies; said to have been invented by Ruth Wakefield of Whitman, Massachusetts, back in 1930.

Conclusion

Certain plants and animals were given by God to humans. These plants and animals were here before humans arrived. In other words, we have been eating from a table set by God. Early food was very simple and close to the earth. Fresh vegetables and meats cooked over a fire to soften them, improve their taste, and stop anything living within them.

Many people today are beginning to return to the old ways – where God started us. Many are starting to buy "organic" foods, grown in a simple way without harmful pesticides. Therefore, new meanings may be given to the words: "Look at the birds ... your heavenly Father feeds them!" (Mt 6: 26) The truth is ... God also continues to feed us!

58. Clothing ...
protection from weather

Why are you anxious concerning clothes?
Learn from the lilies of the field.
They labor not, nor do they spin [thread].
Yet I [Jesus] say to you, that
[King] Solomon in all his splendor was
not clothed (as well) as one of these.

Matthew 6:28-29

The above text says that God takes care of both the wildflowers and we humans. He certainly has given us a start! The skin and hair of many wild animals (especially goats and sheep) and the fibers of many wild plants, including cotton and flax are evidence of his work. Clothes protect humans from the elements, insulating the body from both cold and hot conditions. They importantly act as a barrier between the skin and the environment. They protect the skin during hazardous activities like cooking, and from rough surfaces, splinters, insect bites, rash-causing plants, thorns, infectious or toxic materials, and UV radiation.

Early human cultures, like peoples around the Arctic Circle today, made clothes entirely out of skins and fur. Later, more advanced cultures, replaced the skins with cloth: woven, knitted, or twined; and made from various animal and plant fibers, including flax (to make linen), wool, cotton, silk, and hemp. Many people still wear large sections of cloth wrapped and tied around the body to fit – for example, the "Sari" of women in India.

The earliest evidence of humans using the wild flax plant as a textile comes from the Republic of Georgia, where spun and dyed flax fibers were found in Dzudzuana Cave dating to about 34,000 years ago. Bone "sewing needles" from Kostenki in Russia date to 28,000 years ago. Woven clothes appear on a clay piece found in the Czech Republic dating to 27,000 years ago. By 5000 BC, sheep's hair, or wool, was being spun to make clothes in West Asia, and cotton was being woven into cloth in Egypt's Nile River Valley by 3000 BC.

Weaving is a method of cloth making in which two sets of thread are interlaced at right angles to form a fabric or cloth. The up and down threads are called "warp," and the across threads are called "weft." Weaving requires a "loom" to position the threads. Weaving has been widely used since about 6000 BC.

Knitting uses two or more needles to loop yarn into a series of interconnected loops that will make a garment. Unlike weaving, knitting does not require a loom, making it a valuable technique for peoples who are on the move. The oldest known knitted clothing is a pair of socks from Egypt, dating to about 1000 AD.

Cotton and wool were originally spun into thread by hand. Then, between 500 and 1000 AD, the "spinning wheel" appeared in India. The spinning wheel turns, or spins, plant, or animal fibers into thread. It greatly speeds up the process. Then thread can be woven into cloth. The earliest illustrations of the spinning wheel come from Baghdad (1234 AD), China (1270), and Europe (1280).

In 1733, John Kay in England invented a loom that doubled the speed of weaving. It used a "flying shuttle" on a track with wheels. While the up and down threads were being held tight in traction, the across thread was carried by the shuttle on tracts back and forth. It was thrown, rather than worked through by hand.

In 1764, James Hargreaves in England invented the "Spinning Jenney" which allowed one worker to make thread by turning eight spinning machines at one time.

Then, in 1768, Richard Arkwright invented the "Water Frame," a spinning operation that replaced the worker with a water wheel turned by running water from a stream.

In 1775, James Watt in Scotland invented the rotary steam engine. This engine's steam mechanism <u>turned a wheel</u> that would power machines. In 1779, Samuel Crompton invented the "Spinning Mule" which used steam power to turn over 1000 spinning machines at one time.

In 1785, Edmund Cartwright in England invented the "Power Loom" to weave cloth automatically. By 1818, Manchester, England had 32 cloth making factories containing 5,732 looms. Thus, began the Industrial Revolution.

In 1832, Walter Hunt of New York invented a lock-stitch sewing machine, and in 1849 the safety pin. The lock stitch sewing machine required two threads, one passing through a loop in the other, both interlocking to form part of a seam. Hunt's machine had a needle with an eye at the point and a shuttle operating beneath the cloth to form the lockstitch. Walter Hunt, however, did not patent his machine. Then, in 1846, Elias Howe of the USA built a similar lock-stitch sewing machine with improvements. He was awarded the patent, and many think Howe's was the first practical machine for sewing.

Early in human history, clothes were washed in a nearby river or stream, rubbing them on a sandpaper-like rock and drying them in the sun. The earliest evidence of a soap-like material was found in the excavation of Babylon (in today's Iraq) dating to 2800 BC. Then, in 2200 BC, a formula for soap making was written on a Babylonian clay tablet, consisting of water, alkali, and cassia oil. The Ebers Papyrus of

1550 BC also discusses how ancient Egyptians made soap and bathed regularly. Furthermore, the Bible mentions soap in two verses.

Though you wash yourself with potash and make soap for yourself, your sin is [still] a stain before me, says the Lord Yahweh.

Jeremiah 2:22

He [God] is like a refiner's fire and like fuller's soap.

Malachi 3:2

Soap works by making insoluble dirt and oil particles soluble in water, so that they can be rinsed away. Early soap was probably discovered within the common cooking fire. Fats from cooked meat were mixed with burnt plant ashes (containing potassium hydroxide or lye). This formed a soapy semi-solid brown mass that was found to be useful in cleaning.

In the 600s AD, soap makers were active in Spain and Italy making soap from goat fat and Beech tree ashes. In France, they were using olive oil.

About 1915 AD, German scientists invented a new kind of synthetic soap with the name "detergent." It could wash clothes in cold water without leaving a residue ring.

In 1907, the first electric powered washing machine was invented by Alva F. Fisher in the USA. It used a drum type galvanized tub and an electric motor. An internal agitator moved the clothes around in soapy water, and rollers on top of the machine were used to wring out the water from each piece of clothing. The Bendix Corporation in the USA patented the first fully automatic washer in 1937.

J. Ross Moore in the USA began working on a clothes dryer design in the early 1900s. Finally, after many tries, Hamilton Manufacturing in Wisconsin produced the automatic clothes dryer in 1938.

59. Shelter …
protection from cold and animals

… a thoughtful man [Jesus] built his house on the rock [Peter].
The rain fell, the floods came, and
the winds blew and beat on that house,
but it did not fall, because
it had been founded on rock.

Matthew 7:24-25

By 8,000 BC, people in the Middle East, had begun farming. They no longer moved about, living in tent-like structures. They tended their crops and animals and cooked food in clay ovens. They also built houses.

One of the earliest continuous settlements on this planet was Jericho, located within the present-day country of Israel. Archaeological excavations have dated Jericho's origins to 9,000 BC. The people in Jericho, in its early days, were building houses made with sun-dried bricks - usually with a few weeds in them for rigidity. By 7,000 BC, they were also using mortar to plaster walls and floors.

Catal Huyuk, in modern day Turkey, was one of the earliest "towns" on Earth. By 6,500 BC, it had an estimated population of 6,000 people. Houses in this town were also made of mud bricks.

The earliest house made of timber has been excavated in England. It dates to about 8,000 BC. By 4,000 BC, farming had spread across Europe and the log house became widely used. Logs were laid one upon the other and notched and interlocked at the corners. A stone or bronze ax was used to do these jobs. The gaps between logs were filled in with wood chips and mud.

Originally, forests covered much of the earth's surface and should be considered a gift from God.

God said: 'Behold, I have given to you every plant seed which is on the face of all the Earth, and every tree …
                                                        Genesis 1:29

God made every tree pleasant to the sight and good for food.
                                                        Genesis 2:9

The trees of Yahweh [God the Father] are full;
the cedars of Lebanon that He planted ...

Psalm 104:16

Hiran [king of Tyre] gave Solomon cedar trees
and fir trees [to build the Temple in Jerusalem] ...

1 Kings 5:10

The mountains and hills shall break out into song before
You [God], and all the trees of the field shall clap their
hands.

Isaiah 55:12

(1) Foundation

If a house is to last, it has to have a good foundation. A
foundation made of wood will soon rot when placed in contact with the soil. A rock or cement foundation is more stable and long lasting. Most homes today use cement blocks
or a cement slab foundation.

Cement was used in the ancient world without
knowledge of its chemistry. The Romans discovered how to
make cement by mixing lime and volcanic ash. Lime is calcium oxide, which is obtained by heating limestone in water. The Romans built many cement structures, including
the Pantheon in Rome, which has a 130 foot-wide dome
made with poured concrete.

Modern cement recipes began in 1824 when Joseph Aspdin of England took out a patent for "Portland Cement." He produced his product by firing finely ground clay and limestone. However, his product was fired at too low a temperature. This problem was corrected by Isaac Johnson in 1845. Johnson heated the clay and limestone mixture to 2,642 degrees F. At this temperature, clinkering occurs and its ingredients become strongly cementitious. The material was then ground up into a fine powder, and when mixed with water, resembled "liquid stone" that eventually dried hard.

The Romans tried unsuccessfully to use bronze to reinforce concrete. Then, in 1830, the idea of iron reinforced concrete was suggested in the "Encyclopedia of Cottage, Farm, and Village Architecture." It was not, however, until an 1854 patent by William Wilkinson in England that reinforced concrete became widely used in buildings and bridges.

(2) Framing

Before 1830 heavy timber framing was very common. Vertical supports were few in number and heavy. Walls were constructed into a post and beam skeleton and then filled in with slabs of wood. However, beginning about 1830, a lighter "balloon framing" method became the new normal. In this method lighter, standardized lumber dimensions became available. These were called "studs," or 2 x 4s, and were made possible by new power saws that could cut trees into precise sizes. Less lumber was used and therefore costs went down. Balloon framing also allowed for a large variety of architectural styles.

(3) Plumbing

A good world map will demonstrate that most cities and towns are located along rivers or streams. The reason is that people need fresh water for cooking, bathing, and to flush their waste products away. In rural areas, wells were dug to reach underground water supplies. However, in large cities, outside water sources had to be sought. Ancient Rome, for example, with a million residents, had eleven "aqueducts." These freshwater channels brought millions of gallons of water into Rome each day from as far as fifty miles away from the city. They were powered merely by the force of gravity!

Human waste products have been a difficult problem to solve. These materials provide a good environment for bacteria and a variety of microorganisms to grow. While most microorganisms are the friends of humans, some are not. Germs such as cholera and typhoid are a threat to human life. Therefore, the proper disposal of human waste is an important concern of plumbing within the house.

For thousands of years, the "chamber pot" has been used within buildings to relieve individuals. A "commode" enclosed the chamber pot and provided a surface to sit upon when using it. These "pots" were made first of clay and later of copper. The contents of the pots would often be transferred to a hole in the ground outside a specific building or to a source of nearby running water. Larger buildings might have a "cesspool" nearby that was periodically emptied for use as a fertilizer for farm plants.

The first "flushing toilet" was invented by John Harrington near Bath, England, in 1596 AD. It was installed in the royal household and a few other places. Harrington's toilet included a valve to seal off the toilet and a tank of water to flush away the waste products – features still in use today. However, this early toilet was inadequately vented. Sewer gas constantly leaked into nearby rooms and was a dangerous source of harmful bacteria and a possible explosive force.

In 1775, Alexander Cummings of Scotland invented the S-shaped "trap" to solve the problem of sewer gas. His toilet used "standing water" in the S-shaped trap to seal off the outlet to the bowl. Thus, the foul air, or sewer gas, would not escape back into the room.

Before 1800, most people around the world lived on farms or in small villages. After 1800, towns and cities spread across the planet. The problem of human waste grew larger with the increasing population. The importance of the toilet followed, as did underground water pipes and sewer systems for large population centers. The first underground sewer was installed in New York City in 1728, and the first "water main" carrying fresh water in 1830.

In 1861, Thomas Crapper in London, England, further improved the toilet. He used a pull chain with a large source of water above the toilet for powerful flushing; made an airtight seal between the toilet and the floor; vented its sewer gas with a pipe through the roof; and joined a friend to build pottery-style toilets.

In 1857, single sheets of toilet paper went on sale in the USA. Toilet paper "rolls" were not marketed until 1890.

The first tank-type gas water heater was invented in 1889 by Edwin Ruud who had immigrated from Norway to Pittsburg, Pennsylvania. Hot water, carried by pipes, made the use of the bathtub easier and ushered in the use of the shower to clean oneself.

The Romans used lead pipes because at 621 degrees F. lead melts and can be poured into a pipe-shaped mold. By 1455 AD, German craftsmen had learned how to build hot enough fires to melt iron at 2,800 degrees F. They then poured this metal into castings to make iron pipes. Iron, unlike lead, does not build up in the human body. Copper pipe is also a high temperature metal, melting at 1984 degrees F. Plastic pipe became widely used only after 1966 within the walls of houses.

(4) Electricity

In today's world, most houses are electrically wired. However, in 1850 no homes had electric wires placed within their walls. Later, starting in the early 1900s, each room in a typical house was wired for one overhead light and one wall plug. Today, however, with many household appliances, more outlets are required. In the kitchen there may be the refrigerator, microwave, toaster, blender, electric stove, and dishwasher. In the utility room, the washer, dryer, vacuum cleaner, and sewing machine are used. Bathrooms have hair dryers, the electric toothbrush, and a heating pad. In the work room, the computer and cell phone need recharging. And the air conditioner and heating systems utilize electric motors. Lastly, electric lights are placed throughout the house!

Four important inventions predated the above development. First, the practical DC or direct current generator and motor were developed by Zenobe Gramme of Belgium in 1873. It was called the "Gramme Machine." It generated much higher voltages than earlier generators and could be used in reverse as a DC motor. It was the first powerful electric motor successfully marketed.

Second, in 1879, Thomas Edison developed the first successful electric light bulb. It was based upon the discovery that a bamboo filament in a totally evacuated glass bulb would burn 1200 hours without failing. In 1882, Edison also built a DC generator in New York City and put electric lights in 900 buildings.

Third, Nikola Tesla was born in Croatia in 1856. He is remembered for inventing the AC, or alternating current, electric motor. In 1882, he was walking in a Budapest park and envisioned an iron rotor spinning inside a rotating magnetic field produced by two currents of electricity out of step with one another. In 1883, he built his first AC motor and saw it run. The next year, he immigrated to New York City to work. In 1887, Tesla revealed a commercial version of the AC motor, together with a related AC power generator. In May 1888, he patented this AC generator and motor, and then in July sold his patents to George Westinghouse. In 1895, Tesla and Westinghouse built the first AC power station at Niagara Falls in New York State. Unlike DC power, the AC system could be "stepped up" for transmission to Buffalo, New York (17 miles away), and then "stepped down" for low voltage usage in various homes.

God's invisible force of electricity has thereby been applied to our lives to make the journey to Heaven a little easier.

60. Symphonic Music …
the sounds of Heaven

The Bible witnesses to the importance of music in the spiritual lives of humans.

They entered Jerusalem with harps and with lyres, and
with trumpets to the house of Yahweh [God the Father].

2 Chronicles 20:28

Shout joyfully to Yahweh all the earth …
Sing praise to Yahweh with the lyre …

Psalm 98:4-5

Praise God in his holy place,
praise him in the expanse of his might …
Praise him with the sound of the trumpet;
praise him with the harp and lyre …
Praise him with strings and pipes [flutes] …
Praise him on the cymbals …
Let everything that breathes praise him.

Psalm 150:1-6

For the trumpet will sound, and
the dead will be raised imperishable,
and we will [all] be changed.

1 Corinthians 15:52

The symphony orchestra has a long history. This section will trace the origin of a few important musical instruments and various musical compositions.

The oldest confirmed musical instrument discovered so far is the "flute." In 2008, a 35,000-year old flute was discovered in a cave near Ulm, Germany. It was a five-hole flute with a v-shaped mouthpiece made from a vulture wing bone. A second instrument, an 8-inch three-hole flute, made from a mammoth elephant tusk, dating to between 30,000-37,000 years ago, was also found near Ulm, Germany. As an instrument, the flute produces its sound by the flow of air across an opening.

Another instrument of ancient origin is the "drum." Drums may date back to a time when people first learned how to keep "rhythm." Drums consist of a membrane stretched over a shell and are struck either by the hand or with a drumstick. Ancient drums have been found in China that date to 5,500 BC. The ancient drum, however, had an indefinite pitch. Modern orchestral drums, the "timpani," have a definite pitch. They began to appear in the 1400s AD in Europe but were not much used in music before the 1600s.

The "harp" is a stringed instrument that is plucked with the fingers. Several ancient harps have been found in the royal tomb within the Sumerian town of Ur that date to 3,500 BC. Harps are also depicted on wall paintings in Egyptian tombs dating to 3,000 BC.

The ancient hand-held harp was much smaller than the "pedal harp" of present-day orchestras. The pedal control alters the pitch of the modern harp's 47 strings. The pedal control was invented by Sebastien Erard of France in 1810.

The ancient "trumpet" dates back to 1500 BC. Silver and bronze trumpets were found in King Tut's tomb in Egypt, dating to 1339 BC. These trumpets consisted of a 22-inch straight tube with a bell at the end. The "folded trumpet," without valves, appeared about 1400 AD (after Christ). As with all "brass instruments" the player's lips vibrate, causing the air column to vibrate, producing sound. The modern "piston-valve trumpet" was invented in 1838 by Francois Perinet in France. Valves are used to change the length of the trumpet's tubing. The air is diverted through additional tubing which lowers the tone of the instrument.

The "slide trombone" was invented in the 1400s AD. It uses a telescoping slide to vary the length of the tube which changes the instrument's pitch, going up or down. The trombone's pitch is lower in general than that of the trumpet. The earliest depiction of a trombone occurs in an Italian church painting of around 1490 AD.

The "violin" family consists of the violin (with the highest pitch), then the viola, cello, and double bass (with the lowest pitch). The violin has four strings and a horsehair bow to create the appropriate vibrations. Andrea Amati of Cremona, Italy, was the first known violin maker. Documentation exists for two violins he created in 1542 and 1546. However, these instruments had only three strings.

The first four string violin was also made by A. Amati in 1555 AD.

The "oboe" is a woodwind instrument about 25 inches long with a double reed mouthpiece, metal keys, a conical shape and a flared bell. Sound is created by blowing into the double reed and vibrating its column of air. The oboe was invented about 1650 by the French musicians Jean Hotteterre and Michel Philidos.

The "bassoon" is a double reed woodwind instrument with a distinctive, burping sound. It plays music written in the bass and tenor clefs. It was invented by Martin Hotteterre about 1655 AD.

The "clarinet" is a cylindrical, single-reed instrument. It has many metal keys that can make a variety of fluid sounds. It was invented around 1690 by Johann and Jacob Denner in Germany.

The "piano" is a stringed instrument with a hammer action. It is a highly versatile instrument, capable of playing a wide range of music. Its 88 keys also cover the full range of sounds that the human ear is capable of hearing. The invention of the piano is credited to Bartolomeo Cristofori of Padua, Italy, about 1700. Cristofori was given a position in 1688 by Ferdinando de Medici, himself an excellent musician and music patron.

The "French horn" is a modern instrument, adopted from the trendy sport of hunting. The French horn is a trumpet-like instrument with a larger bell and deeper sound than the trumpet. French hunters in the 1600s developed various versions of this horn. In the 1700s, the "trompe de chasse" looked very much like a modern French horn without the valves, and composers began to write music for this instrument. Then, in 1839 Francois Perinet invented piston valves for the French horn to widen its range.

An important aspect of symphonic music has been the work of its "composers." Here are a few of the heavenly compositions that this group of innovators has given the world.

Year	Composer	Composition
1741	Handel	Messiah
1785	Mozart	Piano Concertos No. 20 and 21
1788	Mozart	Jupiter Symphony (No. 41)
1795	Haydn	London Symphony (No.104)
1804	Beethoven	Symphony No. 3 (Eroica)
1808	Beethoven	Symphony No. 5
1809	Beethoven	Piano Concerto No. 5 ("Emperor")
1846	Mendelssohn	Elijah (Be Not Afraid)
1869	Tchaikovsky	"Romeo and Juliet" Fantasy Overture
1880	Brahms	"Academic Festival" Overture
1893	Dvorak	"New World" Symphony
1901	Rachmaninoff	Piano Concerto No. 2
1901	Elgar	Pomp and Circumstance No. 1
1910	Stravinsky	Firebird Suite

1916	Holst	The Planets ("Jupiter")
1922	Ravel & Mussorgsky	Pictures at an Exhibition
1925	Gershwin	Concerto In F
1928	Gershwin	An American In Paris
1937	Walton	Crown Imperial
1944	Copeland	Appalachian Spring

FINAL THOUGHTS

[1] The Lord Jesus raised various people from the dead. He had power over death, even over his own death. That same power is extended in time by hospitals with doctors and nurses. The Church has always had the Lord's example firmly in mind. Following the Council of Nicaea in 325 AD, the construction of a hospital in every town where there was a bishop was begun. Medieval hospitals were religious communities with the usual care provided by priests and nuns. In France, they were called "hostels of God."

Raising people from the dead is also the duty of the family. Parents are intended by God to "raise" children from conception to adulthood – and even beyond. The parents' work is to nurture life, developing strong, educated, responsible, productive, and moral souls before God. The family is, therefore, the foundation of society.

[2] The Lord Jesus was not bound in space as we are. This was decidedly true after his resurrection on the first Easter Sunday. For example, in Luke 24:31, the reader is told that the risen Jesus instantaneously "disappeared" before two of his disciples in Emmaus, just after "the breaking of the (Eucharistic) bread." However, even before his resurrection, we see Jesus doing amazing things. For example, we see him in one town (Cana), while curing a soldier's little boy in another town (Capernaum) (John 4:46-53). Cana and Capernaum were separated by about 15 miles. Also, on

the Sea of Galilee, when he and his apostles were 3 to 4 miles out from the shore, the reader is told: "immediately the boat was at the land to which they were going" (John 6:21).

Humans, as images of the Lord (Genesis 1:26), are also moving towards a like dominance of space. Transportation advances have made people progressively less bound in space. Just recently an American company began designing a tube-train system that will possibly travel at 700 miles per hour. And so, the movement towards a Heaven-like existence on earth continues.

[3] The Lord Jesus was also not time-bound, as ordinary mortals are. In fact, Jesus, in his divine Person, had no time. Space and time came into existence at his command, when he created matter in the Big Bang and had to put it somewhere. He, however, existed, before the advent of space and time. One example of his power over time is seen in his statements about his own up-coming death and resurrection. He told his apostles three times about these events, almost as if they had already happened (Matthew 16:21; 17:23; 20:19). For example, when he came down from the Mount of the Transfiguration, he said: "Tell no one of this [event you have just seen] until after the Son of Man is raised from the dead!" [Matthew 17:9]

[4] Mankind, as an image of the Lord, has also made great strides in communication technology in the last 200 years. Beginning with the telegraph, humans have gone all the way to the world-wide Internet. One can now speak

with and be seen by another person on the other side of the planet – have a conversation with them. The Pope can address the whole planet and be seen and heard as he speaks. Mankind is becoming less and less bound in both space and in time. This is a clear sign of Heavenly-life, the life God has intended for all of us to live, after leaving this planet.

"God our Savior wants all men [all people] to be saved."
1 Timothy 2:4

[5] Humans are also creating a "New-World." This world-wide activity is another sign that people are created in God's image. Humans are constructing a more Heaven-like place in which to live while still here on earth. This is prophetic. While we continue to struggle with our own sinfulness, the project of a "new world" moves forward. It is not able to be stopped – even by the evil that exists among us! Heaven is coming. The Lord Jesus said so!

God [the Father] did not send the Son [Jesus] into
the world that he might judge the world, but that
the world might be saved through him.
John 3:17

How great are his signs!
And how mighty are his wonders!
His Kingdom is an everlasting Kingdom, and

his dominion is from generation to generation.

Daniel 4:3

I [Daniel] saw ... [one] like the Son of Man [Jesus] who came with the clouds of the heavens, and he came to the Ancient of Days [God the Father] ... And there was given to him [Jesus] dominion and glory, and a Kingdom, that all peoples, nations, and languages shall serve him. His dominion is an everlasting dominion which shall not pass away and shall not be destroyed.

Daniel 7:13-14

Jesus went throughout Galilee ... announcing the good news of the Kingdom, and healing all illness and all sickness in the people.

Matthew 4:23

Do not worry, saying: 'What shall we eat, or what shall we drink?' or 'What shall we wear?' ... seek first the Kingdom of God and his rightness and all these [other] things will be given to you as well.

Matthew 6:31-32

The Kingdom of Heaven is like a treasure hidden in a field [of the world], which someone found ... and in his joy he goes and sells all that he has and buys that field.

Matthew 13:44

I [Jesus] say to you: There are some standing here who will not taste death until they see the Kingdom of God has come in power ... and [then] he brought them up onto a high hill ... and was transfigured in front of them ... becoming glistening white [in appearance before them] ...

Mark 9:1-3

You will see Abraham and Isaac and Jacob, and all the prophets in the Kingdom of God ...

Luke 13:2

After his suffering he [Jesus] presented himself to them living by many convincing proofs; during forty days [after he had risen] being seen by them and teaching about the Kingdom of God.

Acts 1:3

Witnesses ... to us who ate and drank with him after he rose from the dead."

Acts 10:41

We know that if the tent we live in [this body] is unloosed, we have a building from God, a house not made with hands, eternal in Heaven.

2 Corinthians 5:1

www.ingramcontent.com/pod-product-compliance
Lightning Source LLC
Chambersburg PA
CBHW051949090426
42741CB00008B/1326